Smartphone-Based Real-Time Digital Signal Processing

Third Edition

Synthesis Lectures on Signal Processing

Editor
José Moura, *Carnegie Mellon University*

Synthesis Lectures in Signal Processing publishes 80- to 150-page books on topics of interest to signal processing engineers and researchers. The Lectures exploit in detail a focused topic. They can be at different levels of exposition-from a basic introductory tutorial to an advanced monograph-depending on the subject and the goals of the author. Over time, the Lectures will provide a comprehensive treatment of signal processing. Because of its format, the Lectures will also provide current coverage of signal processing, and existing Lectures will be updated by authors when justified.

Lectures in Signal Processing are open to all relevant areas in signal processing. They will cover theory and theoretical methods, algorithms, performance analysis, and applications. Some Lectures will provide a new look at a well established area or problem, while others will venture into a brand new topic in signal processing. By careful reviewing the manuscripts we will strive for quality both in the Lectures' contents and exposition.

Smartphone-Based Real-Time Digital Signal Processing, Third Edition
Nasser Kehtarnavaz, Abhishek Sehgal, Shane Parris, and Arian Azarang
2020

Anywhere-Anytime Signals and Systems Laboratory: from MATLAB to Smartphones, Third Edition
Nasser Kehtarnavaz, Fatemeh Saki, Adrian Duran, and Arian Azarang
2020

Reconstructive-Free Compressive Vision for Surveillance Applications
Henry Braun, Pavan Turaga, Andreas Spanias, Sameeksha Katoch, Suren Jayasuriya, and Cihan Tepedelenlioglu
2019

Smartphone-Based Real-Time Digital Signal Processing, Second Edition
Nasser Kehtarnavaz, Abhishek Sehgal, Shane Parris
2018

iv

DSP for MATLAB™ and LabVIEW™ I: Fundamentals of Discrete Signal Processing
Forester W. Isen
2008

The Theory of Linear Prediction
P. P. Vaidyanathan
2007

Nonlinear Source Separation
Luis B. Almeida
2006

Spectral Analysis of Signals: The Missing Data Case
Yanwei Wang, Jian Li, and Petre Stoica
2006

Smartphone-Based Real-Time Digital Signal Processing, Third Edition

Nasser Kehtarnavaz, Abhishek Sehgal, Shane Parris, and Arian Azarang

ISBN: 978-3-031-01415-4   paperback
ISBN: 978-3-031-02543-3   ebook
ISBN: 978-3-031-00336-3   hardcover

DOI 10.1007/978-3-031-02543-3

A Publication in the Springer series
SYNTHESIS LECTURES ON SIGNAL PROCESSING

Lecture #19
Series Editor: José Moura, *Carnegie Mellon University*
Series ISSN
Print 1932-1236   Electronic 1932-1694

Smartphone-Based Real-Time Digital Signal Processing

Third Edition

Nasser Kehtarnavaz
University of Texas at Dallas

Abhishek Sehgal
University of Texas at Dallas

Shane Parris
University of Texas at Dallas

Arian Azarang
University of Texas at Dallas

SYNTHESIS LECTURES ON SIGNAL PROCESSING #19

ABSTRACT

Real-time or applied digital signal processing courses are offered as follow-ups to conventional or theory-oriented digital signal processing courses in many engineering programs for the purpose of teaching students the technical know-how for putting signal processing algorithms or theory into practical use. These courses normally involve access to a teaching laboratory that is equipped with hardware boards, in particular DSP boards, together with their supporting software. A number of textbooks have been written discussing how to achieve real-time implementation on these hardware boards. This book discusses how to use smartphones as hardware boards for real-time implementation of signal processing algorithms, thus providing an alternative to the hardware boards that are used in signal processing laboratory courses. The fact that mobile devices, in particular smartphones, have become powerful processing platforms led to the development of this book to enable students to use their own smartphones to run signal processing algorithms in real-time considering that these days nearly all students possess smartphones. Changing the hardware platforms that are currently used in applied or real-time signal processing courses to smartphones creates a truly flexible laboratory experience or environment for students. In addition, it relieves the cost burden associated with using dedicated signal processing boards noting that the software development tools for smartphones are free of charge and are well-maintained by smartphone manufacturers. This book is written in such a way that it can be used as a textbook for real-time or applied digital signal processing courses offered at many universities. Ten lab experiments that are commonly encountered in such courses are covered in the book. It is written primarily for those who are already familiar with signal processing concepts and are interested in their real-time and practical aspects. Similar to existing real-time courses, knowledge of C programming is assumed. This book can also be used as a self-study guide for those who wish to become familiar with signal processing app development on either Android or iOS smartphones/tablets. A zipped file of the codes discussed in the book can be acquired from this third-party website.

KEYWORDS

smartphone-based signal processing, real-time signal processing using smartphones, smartphones as signal processing boards

Contents

Preface

Real-time or applied digital signal processing courses are offered as follow-up courses to conventional or theory-oriented digital signal processing courses in many electrical engineering curricula. The purpose of offering real-time or applied digital signal processing courses is to enable students to bridge the gap between signal processing theory and implementation aspects.

A typical real-time or applied digital signal processing course is normally held within the confines of a teaching laboratory room that is equipped with hardware platforms and the accompanying software for those platforms. The fact that mobile devices, in particular smartphones, have become powerful processing platforms led to the development of this book toward enabling students to use their own smartphones as implementation platforms for running signal processing algorithms as apps considering that these days nearly all students possess smartphones. Changing the hardware platforms that are normally used in real-time applied signal processing courses to smartphones creates a truly flexible (anywhere-anytime) laboratory experience or environment for students. In addition, it relieves the cost burden associated with using dedicated signal processing hardware boards noting that the software development tools for smartphones are free of charge and are well-maintained by smartphone manufacturers.

This book is written in such a way that it can be used as a textbook for real-time or applied digital signal processing courses offered at many universities. Ten lab experiments that are commonly encountered in such courses are covered in the book. It is written primarily for those who are already familiar with signal processing concepts and are interested in their real-time and practical aspects. Similar to existing real-time courses, knowledge of C programming is assumed. This book can also be used as a self-study guide for those who wish to become familiar with signal processing app development on either Android or iOS smartphones/tablets. In this third edition, various updates are made to reflect the newer versions of the software tools used in the first and second editions.

The smartphone-based approach covered in this book eases the constraint of a dedicated signal processing laboratory for the purpose of offering applied or real-time signal processing courses as it provides an anywhere-anytime platform for implementation of signal processing algorithms. A zipped file of the codes discussed in the book can be acquired from this third-party website http://sites.fastspring.com/bookcodes/product/SignalProcessingBookcodesThirdEdition.

As a final note, I would like to thank my co-authors and former/current students Abhishek Sehgal, Shane Parris, and Arian Azarang, for their contributions in the first, second, and third editions.

Nasser Kehtarnavaz
Summer 2020

CHAPTER 1

Introduction

Applied or real-time digital signal processing courses offered at many universities have greatly enhanced students' learning of signal processing concepts by covering practical aspects of implementing signal processing algorithms. DSP processor boards are often deployed in these courses. To a lesser extent, ARM-based boards such as Raspberry Pi [1] are utilized. A number of textbooks are available discussing how to implement signal processing algorithms on DSP boards, e.g., [1–6]. This book is written to provide an alternative hardware platform which students can use in an anywhere-anytime manner and at no cost as it is already in their possession, that being their own smartphones.

Not only do there exist hardware and software costs associated with equipping a teaching laboratory with DSP or other types of signal processing boards, in many cases these boards are confined to a specific teaching laboratory location. Taking advantage of the ubiquitous utilization of ARM processors in mobile devices, in particular smartphones, this book covers an alternative approach to teaching applied or real-time DSP courses by enabling students to use their own smartphones to implement signal processing algorithms. Changing the hardware platforms that are currently used in applied or real-time signal processing courses to smartphones creates a truly flexible laboratory experience or environment for students. In addition, it relieves the cost burden associated with using a dedicated signal processing board noting that the software development tools for smartphones are free of charge and are well-maintained.

This book addresses the process of developing signal processing apps on smartphones in a step-by-step manner. It shows how to acquire sensor data, implement typical signal processing algorithms encountered in a real-time or applied digital signal processing course, and how to generate output or display information. It should be noted that these steps are carried out for both the Android and iOS operating systems and besides smartphones, the apps developed can be run on any ARM-based mobile targets such as tablets. The laboratory experiments that are included cover the following topics: signal sampling and i/o buffering, quantization effects, fixed-point vs. floating-point implementation, finite impulse response (FIR) filtering, infinite impulse response (IIR) filtering, adaptive filtering, discrete Fourier transform/fast Fourier transform (DFT/FFT) frequency transformation, and optimization techniques to gain computational efficiency.

1.1   SMARTPHONE IMPLEMENTATION TOOLS

The main challenge in this alternative approach to real-time or applied digital signal processing courses lies in the difference between the programming environments on smartphones and C programming normally used in such real-time courses. Since a typical applied or real-time signal processing course requires familiarity with C programming, the same C programming familiarity is retained for this alternative approach, i.e., by not requiring students to know other programming languages. This challenge is met here by developing Java (for Android smartphones) and Objective-C (for iPhone smartphones) software shells to run C codes seamlessly so that the prerequisite programming knowledge students need to have would be no different than what is currently required.

To allow C codes to be written and compiled on Android smartphone targets, the following cost-free downloadable development tools are utilized: Android Studio [7], Android Software Development Kit (SDK) [7], and Android Native Development Kit (NDK) [8]. These tools provide a comprehensive development environment incorporating an Integrated Development Environment (IDE), Android SDK plug-ins, and an emulator. The NDK provides the support for incorporating C/C++ codes within Android Studio.

To allow C codes to be written and compiled on iOS smartphone targets (iPhones), the Xcode IDE [9] and a Mac computer running OS are utilized. It is worth stating that for iPhone implementation, it is required to register as an iOS developer to be able to run iPhone apps [10]. The Xcode IDE incorporates an editor, the iOS SDK, a built-in debugger for C, and an iPhone/iPad simulator.

1.2   SMARTPHONE IMPLEMENTATION SHELLS

The developed implementation shells for the Android and iOS platforms provide the programming environment needed to perform signal processing laboratory experiments. As shown in Figure 1.1, the shells comprise parts or components which match in functionality for the two platforms. The major difference between the two platforms lies in the programming language. For Android smartphones, the programming language of the shell is Java, and for iOS smartphones, it is Objective-C. Both platforms support implementing codes written in C and this feature is used to provide a uniform programming approach regardless of the type of smartphones students possess.

1.2.1   ANDROID IMPLEMENTATION

The developed Android shell consists of the following three major parts or components.

User Interface   The user interface (UI) comprises the so-called main activity in Java which allows controlling the shell operation and displaying outputs. The component *PreferencesUI* complements the main activity by controlling operational parameters.

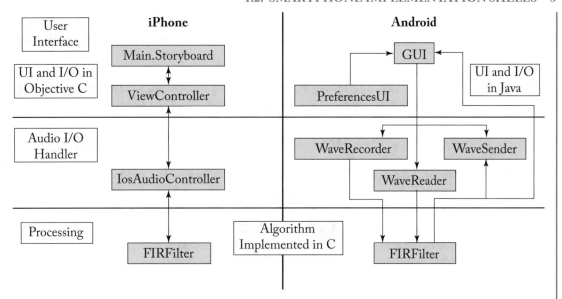

Figure 1.1: Components of the developed shell programs to run C codes on iOS and Android smartphones/tablets.

I/O Handler   The audio input/output (I/O) is split into three modules depending on their functionality. Microphone recording is handled by the module *WaveRecorder*, audio file reading is handled by the module *WaveReader*, and speaker and debug outputs are handled by the module *WaveSaver*. The user is given the option to select one of the two modules *WaveRecorder* and *WaveReader*. In both cases, the module *WaveRecorder* is used for outputs.

Processing   This module allows running C codes within the Android shell. Additional code segments are written to interface with the Java modules using the Java Native Interface (JNI) programming framework.

1.2.2   iOS IMPLEMENTATION

The developed iOS shell also consists of three major parts that match the Android shell:

User Interface (UI)   The UI module handles displaying program outputs and all interactions with the user. It provides an interface for the user to change the parameters required by the processing algorithm. The component *Main.Storyboard* contains the UI elements and the component *ViewController* handles all UI events such as user interaction or parameter changes.

I/O Handler   This module gathers data to be processed by the processing algorithm depending on the source specified by the user. The component *IosAudioController* described in [11] is used to gather data from the microphone and provide data to a C code for processing. This component is also responsible for outputting processed audio signals to the speaker. When using an audio file, the component *audioReader* handles reading stored audio files and passing them to C codes for processing.

Processing   This module allows running signal processing algorithms for the lab experiments that are written in C. This module processes and returns data provided by the I/O Handler.

1.3   OVERVIEW OF ARM PROCESSOR ARCHITECTURE

ARM is the processing engine that is used in modern smartphones. The ARM (Advanced RISC Machine) architecture has been extensively used in embedded systems. Its designs are licensed and incorporated into a wide range of embedded systems and low power mobile devices. The ARM architecture refers to a family of reduced instruction set computing (RISC) architectures produced by the company ARM. The most common architectures currently in use for mobile devices are the ARMv7 architecture which supports 32-bit addressing/arithmetic and the ARMv8 architecture which supports 64-bit addressing/arithmetic. An overview of the ARMv7 architecture is provided next.

1.3.1   DATA FLOW AND REGISTERS

The RISC nature of the ARM architecture means that arithmetic operations take place in a load/store manner. Figure 1.2 shows a diagram of the dataflow in an ARM core. ARM registers, that are all of uniform 32-bit width, consist of 13 general purpose registers (r0 to r12) and these three additional special use registers: stack pointer (SP or r13) which contains a pointer to the active stack, link register (LR or r14) which stores a return value when a branch instruction is called, and program counter (PC or r15) which contains a pointer to the current instruction being executed. In addition, there is one special register called Current Program Status Register (CPSR) which holds Application Program Status Register (APSR) and additional processor state flags. APSR refers to the ALU status flag bits set by the previous instruction in bits 31 to 27 of CPSR. Starting with bit 31, these values indicate negative, zero, carry, overflow, and saturation.

The execution pipeline varies between different versions of the ARM architecture. Instructions can be either from the ARM instruction set, which consists of 32-bit instructions, or from the Thumb instruction set, which consists of 16-bit instructions providing a compact data processing capability.

Some other features of the ARM architecture include barrel shifter, shown as part of the ALU in Figure 1.3, which is capable of performing logical left and right shifts, arithmetic right shifts, rotate right, and rotate right extended operations on operand B. Another feature is the ability to perform conditional execution. For instance, when decrementing an index as part of a loop, the test for zero can be performed with no overhead as part of the subtraction operation; the condition result is then used to break out of the loop. Other features such as the Advanced

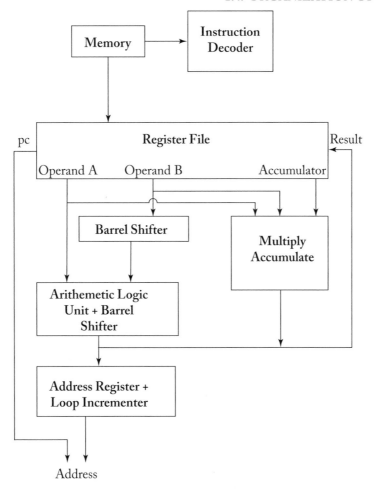

Figure 1.2: ARM processor data flow.

Single Instruction Multiple Data (SIMD) (NEON) coprocessor [12] will be discussed in later chapters. Interested readers can refer to [13] for additional and more detailed materials regarding the ARM architecture.

1.4   ORGANIZATION OF CHAPTERS

The chapters that follow are organized as follows. In Chapters 2 and 3, the smartphone software tools are presented, and the steps one needs to take in order to create a basic smartphone app are discussed. Chapter 2 covers the setup of the Android Studio programming environment, and Lab L1 shows the development of a "Hello World" app for Android smartphones.

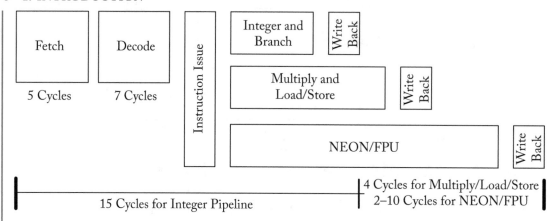

Figure 1.3: ARM cortex-A15 instruction pipeline.

Chapter 3 and Lab L2 are the counterparts of Chapter 2 and Lab L1 focusing instead on the iOS operating system. Chapter 3 details the setup of the Xcode programming environment and duplicates the "Hello World" app from Lab L1. It also includes the debugging tool for iOS smartphones.

Chapter 4 introduces the topics of signal sampling and frame-based processing, and the steps that are required to interface with the A/D and D/A (analog-to-digital and digital-to-analog) converters for audio signal input and output on a smartphone target. As part of this process, the smartphone app shells for the Android and iOS smartphone platforms are covered in detail. The Java and Objective-C shells are discussed, and the steps to incorporate C codes are explained.

Labs L3 and L4 in Chapter 4 show how to sample an analog signal, process it, and produce an output in real-time on an Android and iOS smartphone target, respectively. Lab L3 covers the Android development environment, and Lab L4 the iOS development environment. These lab experiments involve processing a frame of signal samples captured by the smartphone microphone. The frame length can be altered by the user through a graphical-user-interface (GUI) settings menu. The sampling rate can also be altered depending on the sampling rates permitted by the A/D converter of the smartphone target used. It is normally possible to alter the sampling rate on a smartphone from 8–48 kHz. A lowpass FIR filter together with a user-specified delay are considered in this lab experiment. The delay is meant to simulate an additional signal processing algorithm running on the ARM processor of the smartphone. The delay can be changed by the user through the settings menu, adding additional processing time to the lowpass filtering time. By increasing the sampling frequency or lowering the sampling time interval, data frames will get skipped and hence a real-time throughput cannot be met. Besides skipped frames noted on the GUI, one can hear both the original signal and the filtered signal through the speaker of the smartphone and notice the distortion caused by skipped frames due to the real-time demand. Distortion can also be experienced by increasing the processing time delay, thus demonstrating that a real-time throughput is a balance between computational complexity and computation rate. Processing of one frame of data needs to be done in less than $N * dt$ sec in order to achieve a real-time throughput, where N denotes the frame length and dt the sampling time interval. For example, for a sampling rate of 8 kHz and a frame length of 256, the processing needs to be completed within 32 ms in order for all the frames to get processed without any frames getting skipped.

In Chapter 5, fixed-point and floating-point number representations are discussed and their differences are pointed out. Lab L5 in Chapter 5 gives suggestions on how one may cope with the overflow problem. This lab experiment involves running an FIR filter on a smartphone using fixed-point arithmetic. 16 bits are used to quantize the double-precision floating-point filter coefficients generated by a filter design package. Due to quantization, the frequency response of the filter is affected. The quantization word length can be adjusted in the settings menu and the deviation of the frequency response magnitude can be observed in a graph displayed automatically in the user interface. The settings menu allows the user to alter the quantization bits to examine the deviation of the frequency response from the frequency response of the floating-point implementation. In addition, due to quantization, overflows may occur depending on the number of coefficients. This experiment shows how scaling can be used to overcome overflows by scaling down input samples and scaling back up output samples generated by the filter.

Chapters 6 and 7 discuss common filters used in digital signal processing applications. Lab L6 in Chapter 6 covers FIR filtering and Lab L7 in Chapter 7 shows how adaptive filtering can be used to perform system identification. The experiment in Lab L7 exhibits adaptive filtering where an adaptive FIR filter based on the least mean squares (LMS) coefficient update is implemented to match the output of an IIR filter. The error between the output of the adaptive FIR filter and the IIR filter for an input signal is measured and displayed on the smartphone screen in real-time as the app runs. Over time the error between the two outputs converges toward zero. The user can experiment with the rate of convergence by altering the adaptive filter order through the settings menu without needing to recompile the code. As the filter order is increased, it can be observed that the convergence rate also increases. The drawback of increasing the filter order, that is an increase in the processing time, can also be observed. This experiment allows one to see how a tradeoff between convergence rate and real-time throughput can be established.

Chapter 8 covers frequency domain transforms and their implementation using frame-based processing. Lab L8 explores the computational complexity of Fourier transform algorithms and shows the utilization of Fourier transform for solving linear systems. The first part of this lab experiment compares the computational complexity of DFT and FFT by first computing the DFT directly, having the computational complexity of $O(N^2)$, and then via FFT, having the computational complexity of $O(N \log N)$. In the second part of this lab, a filter is implemented in the frequency domain by using Fourier transform three times. Frequency domain filtering is done by complex multiplication between two transformed signals. This approach is observed to be more computationally efficient than convolution when the length of the filter is made long.

Code efficiency issues are addressed in Chapter 9, in which optimization techniques, as well as the use of intrinsics to access hardware features of the ARM processor, are discussed. Lab L9 in this chapter provides a walkthrough of optimization techniques and their impact on a signal processing app. In this lab experiment, the steps one can take to speed up code execution on a smartphone target are covered. These steps include changing compiler settings, writing efficient C code, and using architecture-specific functions for the ARM processor. The FIR filtering (linear convolution) code is used here to show the effects of these steps on the real-time throughput. Compiler options constitute the simplest but an effective optimization step. By changing these options, the compiler produces executable binaries that are either optimized for higher processing speed or for lower memory footprint. After carrying out various compiler optimization options and observing the computational efficiency gains, one can take advantage of the NEON SIMD coprocessor that modern smartphones possess to perform vector data processing. One method of using the NEON coprocessor is the use of NEON intrinsics within C codes. These intrinsics allow access to architecture specific operations such as fused multiply-accumulate, the Newton–Raphson method for division and square root, data format conversions, and saturating arithmetic operations. In other words, many of the architecture specific features of the ARM processor can be accessed by utilizing intrinsic functions within C codes. The initial processing algorithms can be used as a basis for deciding where to utilize intrinsics. In this lab, it is demonstrated that the convolution of two signal sequences can be performed more efficiently by utilizing a vectorized loop via NEON intrinsics.

Chapter 10 presents an optional alternative approach using the MATLAB Coder [14] from the company MathWorks that can be used to rapidly take a signal processing algorithm implemented in MATLAB and transfer it to a smartphone target. The lab experiment covered in the chapter exhibits the setup process for the MATLAB Coder tool provided by Math-Works which allows converting MATLAB functions into C functions. This requires the use of MATLAB version 2016a with the MATLAB Coder included. The experiment discussed in this chapter shows how to convert a MATLAB function into a C function and implement it into an Android or an iOS app.

1.5   SOFTWARE PACKAGE OF LAB CODES

For performing the laboratory experiments, similar to existing real-time or applied digital signal processing courses, familiarity with C programming and MATLAB are assumed. The lab codes can be obtained from the third-party link http://sites.fastspring.com/bookcodes/product/SignalProcessingBookcodesThirdEdition. The lab subfolders found under Android and iOS in the package include all the codes necessary to implement the labs. For the iOS platform, the necessary software is the Xcode IDE which may be installed from the Mac App Store free of charge. In order to deploy and test apps on an iOS device, note that it is first required to enroll in the iOS Developer Program. The shell for the iOS operating system is included as part of the above package so that all the codes can be acquired together in one place.

In the absence of a smartphone target, the simulator can be used to verify code functionality by using data already stored in a data file or from sensors present on the host machine running the simulator. However, when using the simulator, sensor support is often limited and performance is not comparable to that of an actual smartphone target. To be able to process signals in real-time, an actual smartphone target is needed. All the hardware necessary to run the laboratory experiments covered in this book is available on a modern smartphone.

1.6   REFERENCES

[1] https://www.raspberrypi.org/ 1

[2] N. Kehtarnavaz, *Real-Time Digital Signal Processing Based on the TMS320C6000*, Elsevier, 2004. 1

[3] N. Kehtarnavaz, *Digital Signal Processing System Design*, 2nd ed., LabVIEW-Based Hybrid Programming, Academic Press, 2008. 1

[4] T. Welch, C. Wright, and M. Murrow, *Real-Time Digital Signal Processing from MATLAB to C with the TMS320C6x DSPs*, CRC Press, 2011. DOI: 10.1201/9781420057829. 1

[5] S. Kuo and B. Lee, *Real-Time Digital Signal Processors: Implementations, Applications and Experiments with the TMS320C55x*, Wiley, 2001. DOI: 10.1002/0470035528. 1

[6] N. Kehtarnavaz and S. Mahotra, *Digital Signal Processing Laboratory: LabVIEW-Based FPGA Implementation*, Universal Publishers, 2010. 1

[7] http://developer.android.com/sdk/index.html 2

[8] http://developer.android.com/tools/sdk/ndk/index.html 2

[9] https://developer.apple.com/library/ios/referencelibrary/GettingStarted/RoadMapiOS/index.html 2

[10] https://developer.apple.com/programs/ios/ 2

[11] https://code.google.com/p/ios-coreaudio-example/ 3

[12] http://www.arm.com/products/processors/technologies/neon.php 5

[13] ARM Ltd., *ARM Architecture Reference Manual ARMv7-A and ARMv7-R Edition*, 2011. http://www.arm.com 5

[14] https://www.mathworks.com/products/matlab-coder.html 8

C H A P T E R   2

Android Software Development Tools

This chapter covers the required steps to install the software tools for the development of C codes on Android smartphones. In Chapter 3, the iOS version for iPhone smartphones is covered.

The Android development environment used here is the IntelliJ IDEA-based Android Studio Bundle (Android Studio). C codes are made available to the Android Java environment through the use of the Java Native Interface (JNI) wrapper. Thus, it is also necessary to install the Android Native Development Kit (NDK). This development kit allows one to write C codes, compile, and debug them on an emulated Android platform or on an actual Android smartphone/tablet.

Screenshots are used to show the steps and configuration options involved in the installation when using the Windows operating system. The same software tools are also available for other operating systems.

2.1 INSTALLATION STEPS

Start by creating a directory where the tools are to be installed. A generic directory of *C:\Android* is used here and the setup is done such that all Android development related files are placed within the *C:\Android* directory.

2.1.1 JAVA JDK

If the Java Development Kit (JDK) is not already installed on your computer or you do not have the latest version, download it from Oracle's website and follow the installation steps indicated by the installer. The latest JDK package at the time of this writing can be found on Oracle's website at:

http://www.oracle.com/technetwork/java/javase/downloads/index.html

Click on the JDK *Download* button in the *Java Platform, Standard Edition* section shown in Figure 2.1a, and you will be taken to the page shown in Figure 2.1b. From the list of supported platforms, select the correct version for your operating system. For example, if you are running a 64-bit operating system, select the appropriate package.

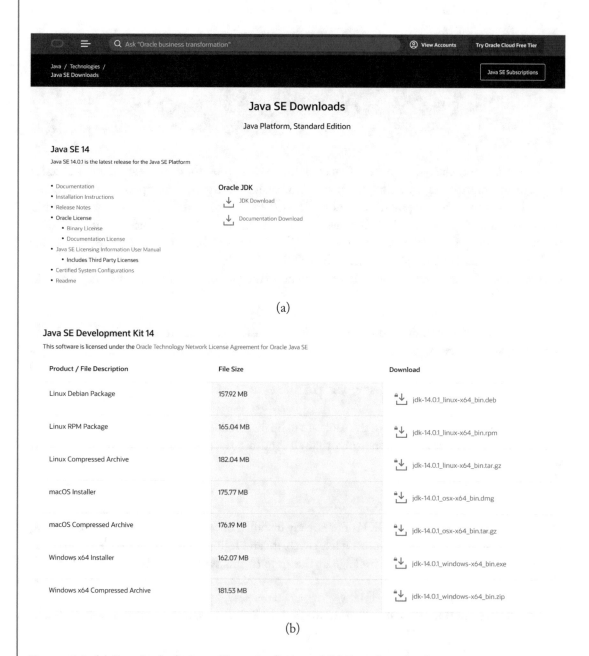

Figure 2.1: (a) Standard edition of Java platform and (b) Java downloads.

Command line tools only

If you do not need Android Studio, you can download the basic Android command line tools below. You can use the included sdkmanager to download other SDK packages.

These tools are included in Android Studio.

Platform	SDK tools package	Size	SHA-256 checksum
Windows	commandlinetools-win-6514223_latest.zip	82 MB	408cbc88f1a077521e20574ecc0a3c45d2d2a9e099011bb0493cb22779811cc3
Mac	commandlinetools-mac-6514223_latest.zip	82 MB	d11cca1138d35852cc786ecb3cc26d0ce4fbfc718a6fb2f369d4bb21af230ce7
Linux	commandlinetools-linux-6514223_latest.zip	82 MB	ef319a5afdb41822cb1c88d93bc7c23b0af4fc670abca89ff0346ee6688da797

Figure 2.2: **SDK packages.**

2.1.2   ANDROID STUDIO BUNDLE AND NATIVE DEVELOPMENT KIT

The most recent versions of Android Studio and the NDK at the time of this writing are used to run the lab experiments in the book. For the Windows installation, the Android Studio is available as an executable installer which incorporates the development environment. The SDK tools need to be installed separately from the development environment. In both cases, the installation binaries appear at this website as depicted in Figure 2.2:

http://developer.android.com/sdk/index.html

The Android NDK is available in the form of a self-extracting archive at this website under the *Download* section:

http://developer.android.com/tools/sdk/ndk/index.html

Download the Android Studio, SDK, and NDK installation binaries into the *Android* directory created earlier and run the Android Studio installer.

During the installation of Android Studio, there are two important settings that are critical to do correctly; see Figures 2.3 and 2.4. For the setting shown in Figure 2.3, make sure that all the components are selected for installation, and for the setting shown in Figure 2.4, make sure that Android Studio is installed in the directory *C:\Android*. The same procedure needs to be done for the Android SDK installation. To do so, manually create the directories by using the Browse option and create a Studio folder and a *sdk* folder. When the installer is finished, do not allow it to start Android Studio as additional configuration is still needed.

The last step is to extract the Android NDK to the folder *C:\Android* by placing the archive executable in the folder and running it. When this action is completed, rename the folder *android-ndk-<version>* to *ndk*.

2.1.3   ENVIRONMENT VARIABLE CONFIGURATION

Before running Android Studio for the first time, the system environment needs to be set up by adding the SDK *platform-tools* folder to the system *path* variable and setting the variables to define the Android Virtual Device (AVD) storage location as well as the locations for the

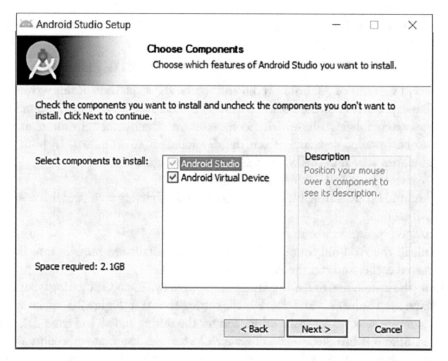

Figure 2.3: Android studio setup.

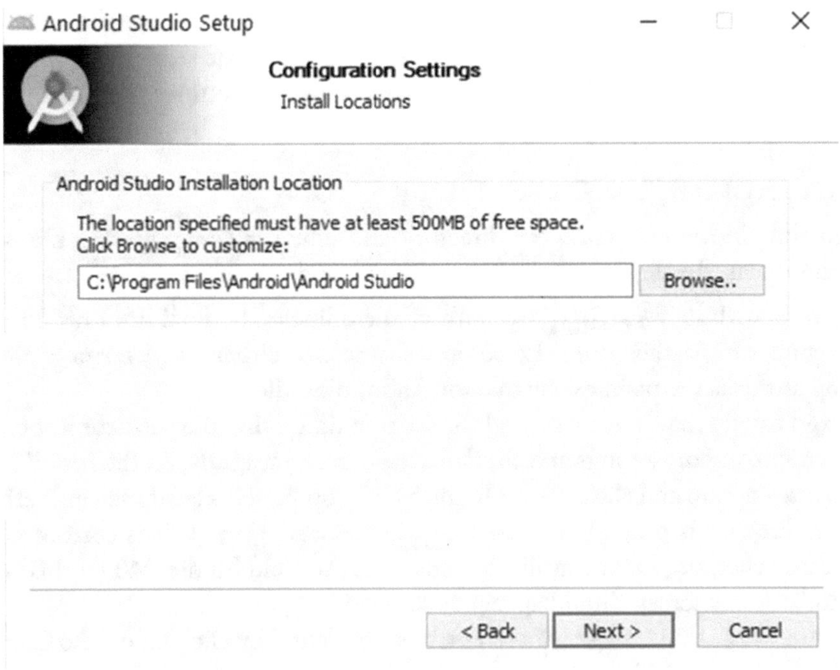

Figure 2.4: Configuration settings.

Android SDK and NDK. The steps listed here are for the Windows operating system. Similar steps need to be followed for other operating systems.

On your desktop, right click on the *Computer* icon and select *Properties*. Next, open the *Advanced system settings* menu and click on *Environment Variables* at the bottom of the *Advanced* tab; see Figure 2.5. Then, create new system variables by clicking the *New...* button below the *System variables* section, shown in Figure 2.6. There are three new system variables that need to be set: `ANDROID_SDK_HOME` with the value `C:\Android`, `ANDROID_SDK_ROOT` with the value `%ANDROID_SDK_HOME%\sdk`, and `ANDROID_NDK_HOME` with the value `%ANDROID_SDK_HOME%\ndk`.

Then, add the following text to the end of your system *path* variable as shown in Figure 2.7; `%ANDROID_SKD_ROOT%\platform-tools`. Be sure to include the semicolon which is used to separate the variables. Modifications are now complete and the settings menus can be closed.

2.1.4   ANDROID STUDIO CONFIGURATION

Navigate to the *C:\Android\Studio\bin* directory and edit the *idea.properties* file with a text editor. Uncomment the lines `idea.config.path` and `idea.system.path` and replace `${idea.home}` with `C:/Android` for both of these lines. The final lines need to appear as shown in Figure 2.8. At this time, also setup a shortcut to either *studio.exe* or *studio64.exe* on your desktop as this is the main executable for Android Studio.

Android Studio may now be started for the first time using the shortcut just created. The tool will prompt to import settings and check for the correct Java JDK. At the *Install Type* screen, select the *Custom* option and click *Next*. On the SDK Components Setup screen, verify that the *Android SDK Location* is properly detected as `C:\Android\sdk`. If it is correct, click *Finish* which will cause checking for any available updates to Android Studio. When this is done, the Android Studio home screen should appear as shown in Figure 2.9.

Now run the *SDK Manager*, whose entry can be found by clicking on the Configure option. The SDK Manager will automatically select any components of the Android SDK which need updating, as illustrated in Figure 2.10. From this menu, additional system images for emulation and API packages for future Android versions can get added. Make sure that CMake, LLDB, and NDK (Side by side) are installed from this page.

Click the Install option and allow the update process to complete.

2.1.5   ANDROID EMULATOR CONFIGURATION

The last item to take care of is configuring an Android Virtual Device (AVD) on which the emulation and debugging are to be performed. From within the SDK Manager, open the *Tools* menu (see Figure 2.10) and select the *Manage AVDs* option to open the Android Virtual Device Manager, as shown in Figure 2.11. By default, Android Studio creates an x86 AVD. Since our

Figure 2.5: **System properties.**

Figure 2.6: **Environment variables.**

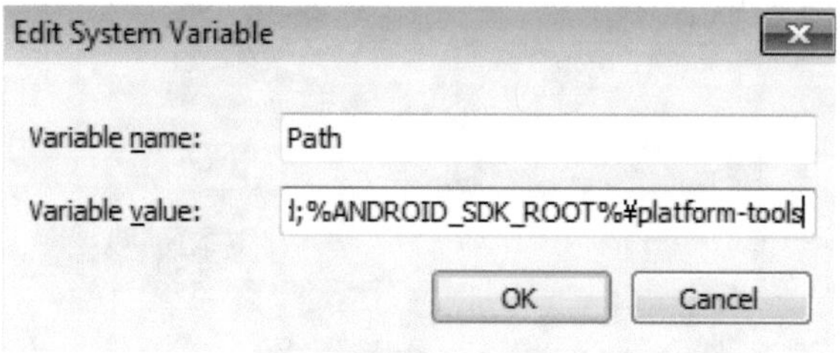

Figure 2.7: Edit system variable.

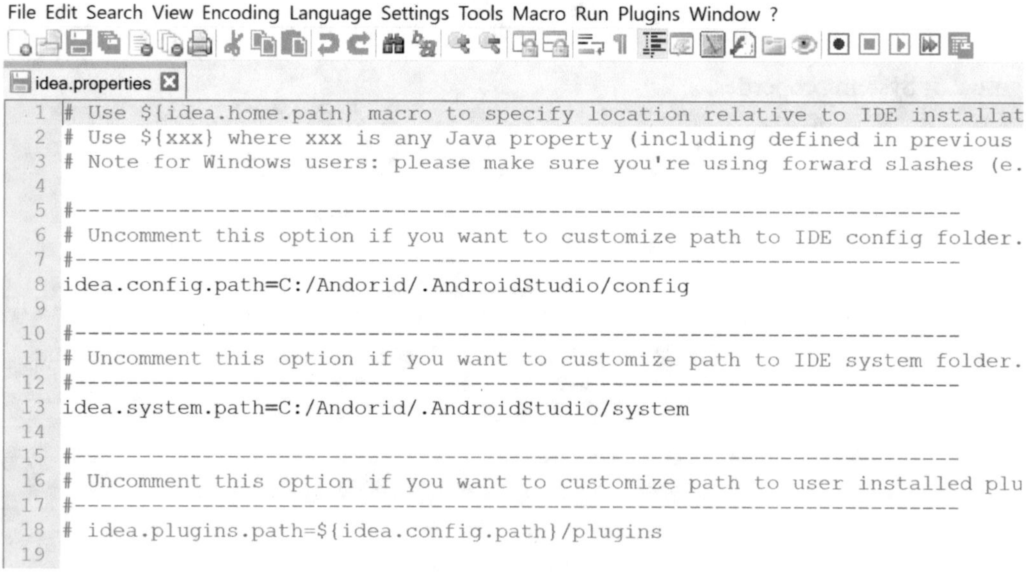

Figure 2.8: idea.properties.

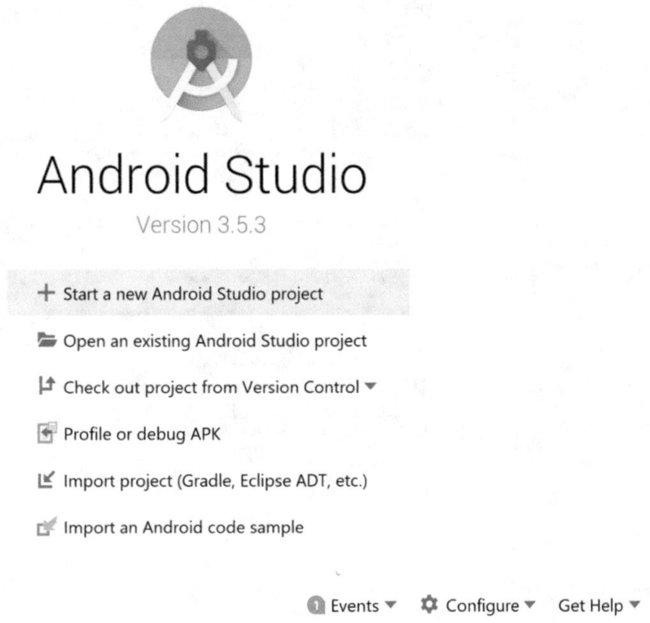

Android Studio

Version 3.5.3

+ Start a new Android Studio project

📁 Open an existing Android Studio project

↳ Check out project from Version Control ▼

⬇ Profile or debug APK

↳ Import project (Gradle, Eclipse ADT, etc.)

↳ Import an Android code sample

❶ Events ▼   ⚙ Configure ▼   Get Help ▼

Figure 2.9: **Android studio.**

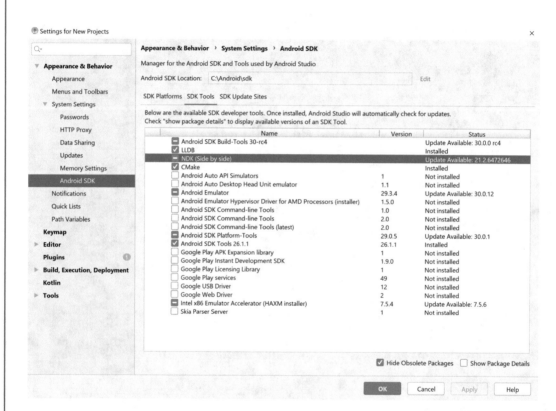

Figure 2.10: **SDK tools.**

Figure 2.11: Manage Android virtual device (AVD).

development focus is on ARM-based implementations, an ARM-based emulator instance is required. Begin by deleting any existing AVD instances.

During the installation of a virtual device, if encountering the error shown in Figure 2.12, enable VT-x in BIOS to remove this error.

1. Click the *Create...* button to start configuring the AVD (see Figure 2.13a).

2. Select the preferred device to use from the list. For the *Device* option, select a device with a good screen resolution for your computer—*Nexus 5X* is normally fine (see Figure 2.13b). On the next page, a list of Android release versions appear. At the *ABI* column, "*Armeabi-v7a*", select ABI. If you are not able to see the option "Armeabi-v7a," switch tab to "Other Images," and download Lollipop Android 5.1 Armeabi-v7a.

For compatibility with smartphones released within the last two years, it is suggested to select the *Target* as the latest available version with the *CPU/ABI* as *ARM (armeabi-v7a)*. You should now be able to create the AVD by clicking *Finish*. Select the AVD you just created in the list of devices (see Figure 2.14). Go to the *AVD Manager* from *Tools*, then select last column *Action*. Click *Launch* and wait for the AVD to boot (see Figure 2.15). Once the AVD launches, unlock the screen and get rid of the greeting message (see Figure 2.16). To run the Android emulator faster, Quick Boot can be used. Note that for the first time, the emulator needs to be run using Cold Boot. By using Quick Boot for subsequent starts, the latest state is stored and used for future starts. This feature is set as default (see Figure 2.17).

2.1.6   ANDROID STUDIO SETUP FOR MAC

For Mac OS installation, the following softwares need to be installed: Xcode, Java Development Kit (JDK), Android Studio Development Bundle, Android Native Development Kit (NDK). Xcode can be downloaded from the following link:

https://developer.apple.com/xcode/download/

Set the environment variable ANDROID_HOME to your Android SDK location. The correct syntax for setting your environment variable is:

```
export ANDROID_HOME=<path to SDK>/android/sdk
```

In order to find the SDK path, launch Android Studio and find the SDK location:

File > Project Structure > SDK Location: "Android SDK location"

The path for the SDK and platform tools directories is:

```
PATH: PATH=$PATH:$ANDROID_HOME/tools:$ANDROID_HOME/platform-tools
```

Once NDK is downloaded, set the NDK location to: `home/dev folder (~/dev)`.

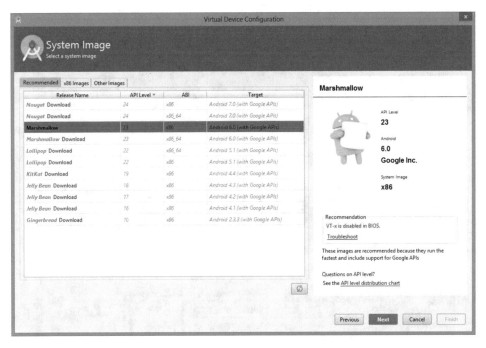

Figure 2.12: VT-x is disabled in BIOS.

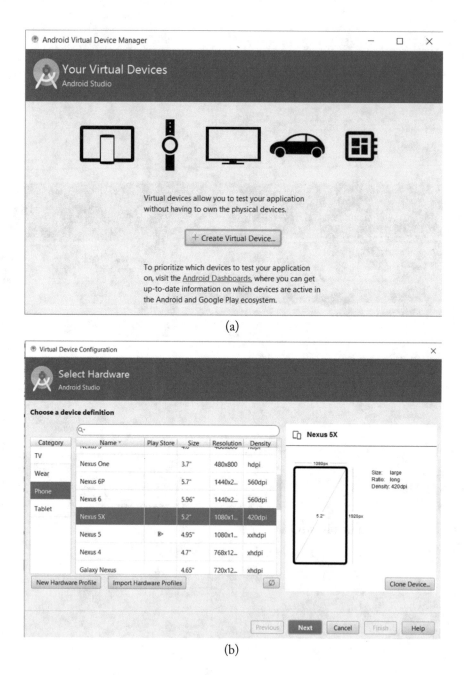

Figure 2.13: (a) AVD and (b) AVD selection.

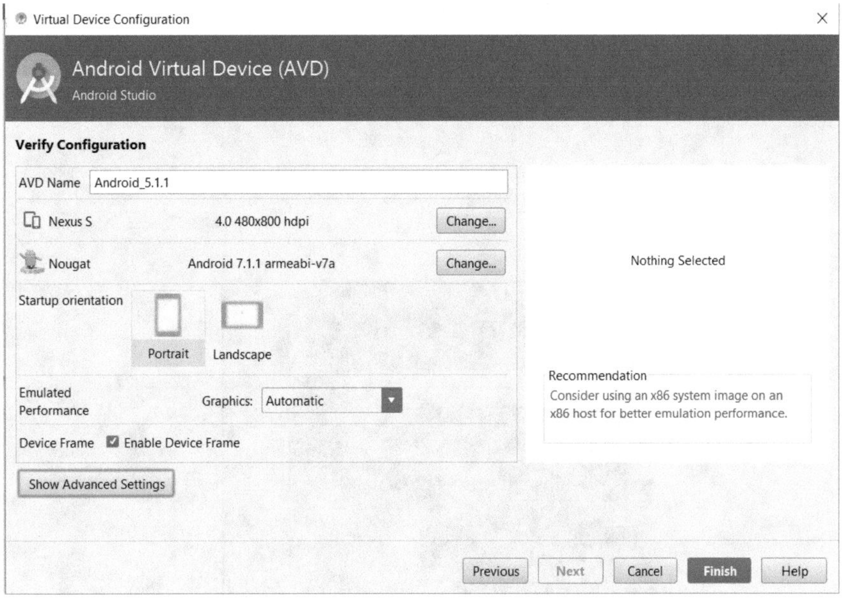

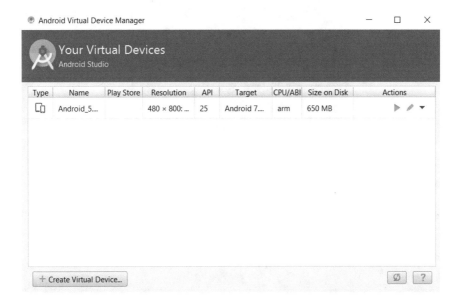

Figure 2.14: **AVD** setting.

Figure 2.15: **AVD** booting.

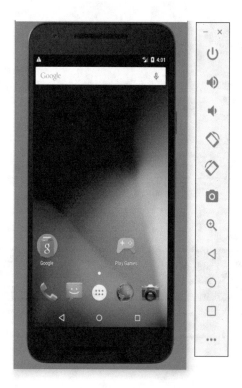

Figure 2.16: **AVD** appearance.

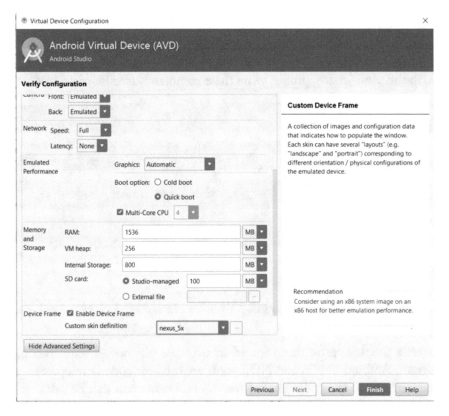

Figure 2.17: Quick boot.

It is worth noting that the Android NDK or SDK folders should not be appear inside each other.

Set the `ANDROID_NDK_HOME` environment variable to the directory where the NDK is installed; for example:

export ANDROID_NDK_HOME=~/dev/android-ndk-r14b

Then the NDK directory should be added to PATH:

export PATH=$PATH:$ANDROID_NDK_HOME

The SDK is usually installed in the library. In order to find the folder, use `cd ~/Library`, and then `open ~/Library/` in the terminal to find full path of SDK.

To check, write the following command in the terminal:

nano ~/.bash_profile

Then, in the nano~/bash_profile, write these commands (see Figure 2.18):

```
export ANDROID_HOME={YOUR_PATH}
```

```
export PATH=$PATH:$ANDROID_HOME/tools:$ANDROID_HOME/platform-tools
```

```
export ANDROID_NDK_HOME=[NDK_PATH]/android-ndk-[version]
```

```
export PATH=$PATH:$ANDROID_NDK_HOME
```

L1   LAB 1: GETTING FAMILIAR WITH ANDROID SOFTWARE TOOLS

This lab covers a simple app on the Android smartphone platform by constructing a "Hello world!" program. Android Studio and NDK tools are used for code development, emulation, and code debugging. All the codes needed for this and other labs can be extracted from the package mentioned in Chapter 1. Start by launching Android Studio, and if not already done, set up an Android Virtual Device (AVD) for use with the Android emulator.

- Begin by creating a new Android project using the Quick Start menu found on the Android Studio home screen.

- Set the *Application Name* to `HelloWorld` and the project location to a folder within the C:\Android directory.

- Change the *Company Domain* to `dsp.com` so that the package name reads as `com.dsp.helloworld`. This is of importance later as it will affect the naming of your native code methods. Refer to Figure 2.19a for the previous three steps.

Figure 2.18: Commands.

- The `Target` Android device should be set to `Phone and Tablet` using a `Minimum SDK` setting of *API 15*.

- Click *Next* and on the following screen choose to create a `Empty Activity`.

- Click *Next* again to create a *Empty Activity* and leave the default naming.

- Select *Finish*. The new app project is now created and the main app editor will open to show the GUI layout of the app.

- Navigate to the *java* directory of the app in the *Project* window and open the *MainActivity.java* file under *com.dsp.helloworld*, as shown in Figure 2.19b.

The class that typically defines an Android app is called an *Activity*. Activities are generally used to define user interface elements. An Android app has activities containing various sections that the user might interact with such as the main app window. Activities can also be used to construct and display other activities - such as if a settings window is needed. Whenever an Android app is opened, the `onCreate` function or method is called. This method can be regarded as the "main" of an activity. Other methods may also be called during various portions of the app lifecycle as detailed at the following website:

http://developer.android.com/training/basics/activity-lifecycle/starting.html

In the default code created by the SDK, `setContentView(R.layout.activity_main)` exhibits the GUI. The layout is described in the file *res/layout/activity_main.xml* in the *Package*

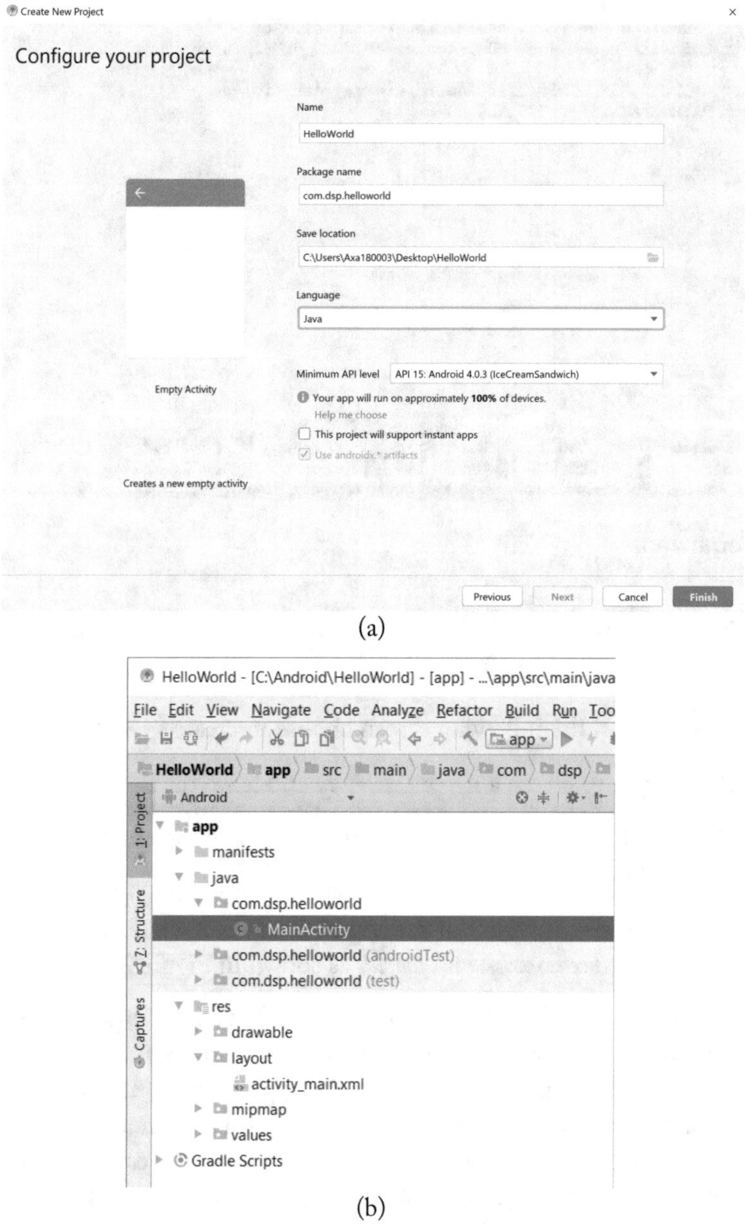

(a)

(b)

Figure 2.19: (a) Create Android project and (b) HelloWorld project.

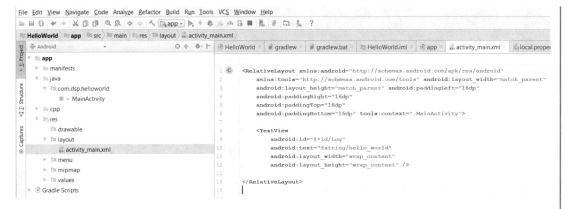

Figure 2.20: XML text.

Explorer window. Open this file to preview the user interface. Layouts can be modified using the WYSIWYG editor which is built into Android Studio. For now the basic GUI suits our purposes with one minor modification detailed as follows:

- Open the XML text of the layout (see Figure 2.20) by double clicking on the `Hello world!` text or by clicking on the *activity_main.xml* tab next to the *Graphical Layout* tab.

- Add the line `android:id=''@+id/Log''` within the `<TextView/>` section on a new line and save the changes. This gives a name to the TextView UI element.

TextView in the GUI acts similar to a console window. It displays text. Additional text can be appended to it. By adding the `android:id` directive to TextView, it may be interfaced with in the app code.

After setting up the emulator and the app GUI, let us now cover interfacing with C codes. Note that it is not required to know the Java code syntax. The purpose is to show that the Java Native Interface (JNI) is a bridge between Java and C codes. Java is useful for handling Android APIs for sound and video i/o, whereas the signal processing codes are done in C. Of course, familiarity with C programming is assumed.

A string returned from a C code is considered here. The procedure to integrate native code consists of creating a C code segment and performing more alterations to the project. First, it is required to add support for the native C code to the project. The first step is to create a folder in which the C code will be stored. In the Project listing, navigate down to New > Folder > JNI to create a folder in the listing called `jni`. Refer to Figures 2.21–2.24. Figure 2.23 shows how the Project listing view may be changed in order to show the jni folder in the main source listing.

Figure 2.21: JNI folder.

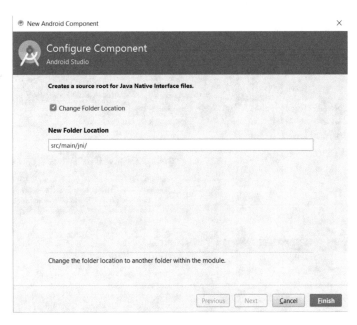

Figure 2.22: Folder location.

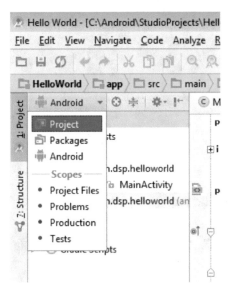

Figure 2.23: Project listing.

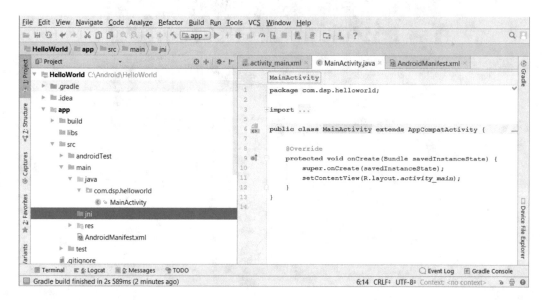

Figure 2.24: MainActivity.java.

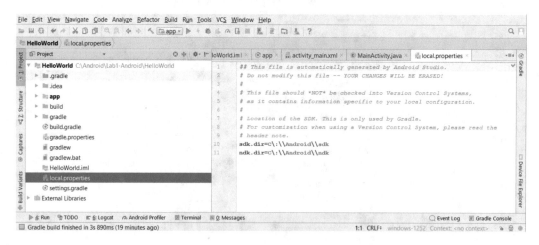

Figure 2.25: local.properties.

Android Studio now needs to be configured to build a C code using the Gradle build system. Begin by specifying the NDK location in the project *local.properties* file according to Figure 2.24. Assuming the directory *C:/Android* is used for setting up the development tools, the location specification would be as follows:

```
ndk.dir=C\:\\Android\\ndk
```

Next, the native library specification needs to get added to the *build.gradle* file within the project listing under the "app" folder. This specification declares the type of external build and its configuration file which defines the name of the external library that Java will load. This is achieved by adding the following code within the *android* section:

```
externalNativeBuild {
    cmake {
        path ''CMakeLists.txt''
    }
}
```

The correct placement of the code is shown in Figure 2.26. Another part that needs to get added is a `CMakeLists.txt` file in the *project* section and the app folder as shown in Figure 2.27 and by having the following code inside the .txt file:

cmake_minimum_required (**VERSION 3.4.1**)
Creates and names a library, sets it as either STATIC
or SHARED, and provides the relative paths to its source code.
You can define multiple libraries, and CMake builds them for you.
Gradle automatically packages shared libraries with your APK.
add_library(# *Sets the name of the library.*
                 HelloWorld

                 # *Sets the library as a shared library.*
                 SHARED

                 # *Provides a relative path to your source file(s).*
                 src/main/jni/HelloWorld.c)

)

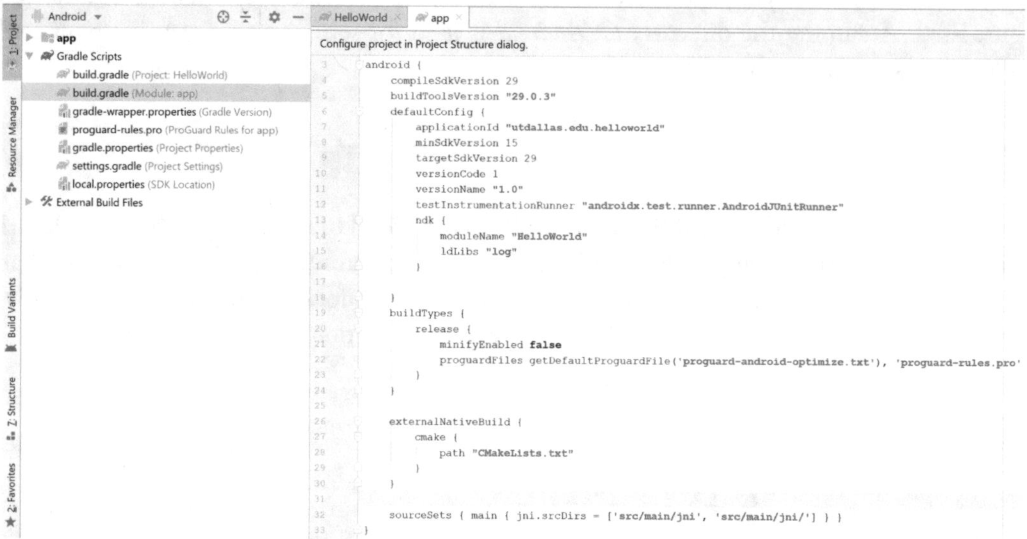

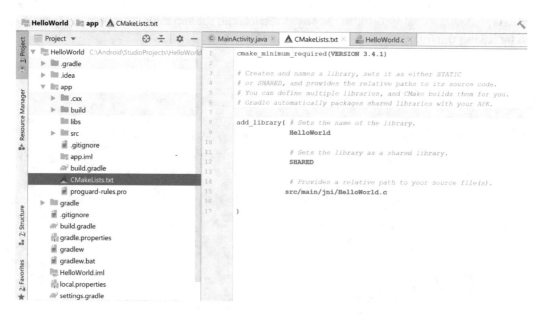

Figure 2.26: **Correct placement of code.**

Figure 2.27: **CMakeLists.txt.**

The C code considered here consists of a simple method to return a string when it is called from the `onCreate` method. First, the code that defines the native method needs to be included. Create a new *HelloWorld.c* file. Add the following code and save the changes.

```
#import <jni.h>

jstring Java_com_dsp_helloworld_MainActivity_getString ( JNIEnv* env,
      jobject thiz )  {
                      return (*env)->NewStringUTF(env, ''Hello UTD!'');
}
```

This code defines a method that returns a Java string object according to the JNI specifications with the text `Hello UTD!`. The naming for this method is dependent on what is called *fully qualified name* of the native method which is defined in the *MainActivity* class. There are alternate methods of defining native methods that will be discussed in later labs.

It is important to note that due to a bug currently present in the Gradle build system, a dummy C source file needs to be created in the jni folder in order for the build process to complete successfully. Simply create a new source file, named dummy.c for example, without any code content.

Next, the native method needs to be declared within the *MainActivity.java* class (see Figure 2.28) according to the naming used in the C code. To do so, add this declaration below the `onCreate` method already defined.

```
public native String getString();
```

Now, add the following code within `public class` to load the native library:

```
static {
      System.loadLibrary(''HelloWorld'');
}
```

To use the TextView GUI object, it needs to be imported by adding the following declaration to the top of the *MainActivity.java* file:

```
import android.widget.TextView;
```

The TextView defined in the GUI layout needs to be hooked to the `onCreate` method by adding the following lines to the end of the onCreate method code section (after `setContentView` but inside the bracket):

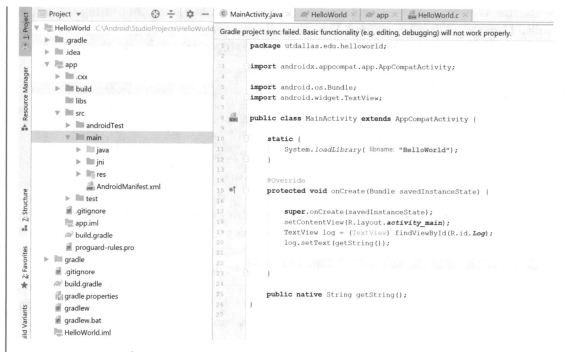

Figure 2.28: **MainActivity** class.

```
TextView log = (TextView)findViewById(R.id.Log);
log.setText( getString() );
```

This will cause the text displayed in the TextView to be changed by the second line which calls the C `getString` method.

Save the changes and select the *Make Project* option (located under the Build category on the main toolbar). Android Studio would display the build progress and notify if any errors occur. Next, run the app on the Android emulator using the *Run app* option located in the Run menu of the toolbar. If an emulator is already running, an option will be given to deploy the app to the selected device (see Figure 2.29). Android Studio should launch the emulator and the screen (see Figure 2.30) would display `Hello UTD!` . To confirm that the display is being changed, comment out the line `log.setText()` and run the app again. This time the screen would display `Hello World!`

Note that the LogCat feature of Android Studio can be used to display a message from the C code. LogCat is equivalent to the main system log or display of the execution information. Here, the code from the previous project is modified to enable the log output capability as follows:

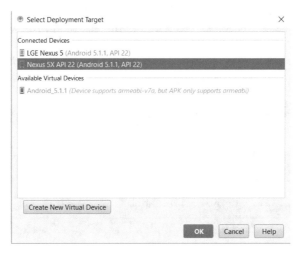

Figure 2.29: Deployment target.

Figure 2.30: Emulator screen.

- Add the logging library to the *build.gradle* file (see Figure 2.26) by adding the line `ldLibs ''log''` to the *ndk* section (add the *ndk* section as well).

- Add the Android logging import to the top of the *HelloWorld.c* source file by adding the line `#include <android/log.h>`.

- Add the following code to output the test message before the `return` statement:

```
int classNum = 9001;
int secNum = 1;
__android_log_print(ANDROID_LOG_ERROR, ''HelloWorld'',
                ''DSP %d.%03d'', classNum, secNum);
```

The `__android_log_print()` method (two underscores at the beginning) is similar to the printf function in C. The first two parameters are the log level and the message tag. The logging level is the priority of the message, the list of which can be found in the *android/log.h* header file. The tag is used to help identify the source of the message; in this case the name of the app. The next parameter is the message to be logged. For the above example, the string has the specified integer for the class number inserted, followed by the specified integer for the section number. The same number formatting that is possible when using the printf function may also be used here. For instance, the section number can be formatted to three characters width with leading zeros. Variables are last and are inserted with the formatting specified in the message string in the order they are listed.

Save the changes made to the *HelloWorld.c* source file and run the app again. This time, Android Studio should automatically open the Android DDMS window and show the LogCat screen. The message `DSP 9001.001` would appear in the listing if the previous procedures were performed properly (see Figure 2.32).

L1.1   LAB EXERCISE

Write a C function within the above Android shell to implement the following difference equation: $y(n) = a * y(n-1) + x(n)$.

Let $x(n)$ be a unit sample at time $n = 0$, and $a = 0.5$. Find and display the output $y(n)$ for n values from 0–20. Explain how the output changes as the coefficient a is varied. Outputs need to be displayed on the main app screen as well as being sent to LogCat using the Log library.

Hints   This exercise addresses the use of JNI conventions for native methods. Using the example above as a template, implement the difference equation as a C function which takes integer

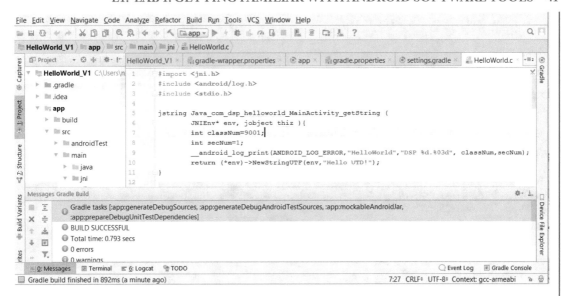

Figure 2.31: HelloWorld.c.

Table 2.1: Data type conversions

Java	JNI	C
double	jdouble	double
float	jfloat	float
long	jlong	long
int	jint	int
short	jshort	short
boolean	jboolean	int
float[]	jfloatarray	float *
float[][]	jobjectarray	float **

input for the variable n and outputs the floating-point result $y(n)$. Use the relations shown in Table 2.1 as a reference for matching data types between Java, JNI, and C.

Table 2.1 shows some common data types. A multi-dimensional array is represented as an array of arrays. With an array being an object in Java, a multi-dimensional array appears as an array of Java object primitives (which are themselves arrays of floating-point primitives).

For the example above, the function used is:

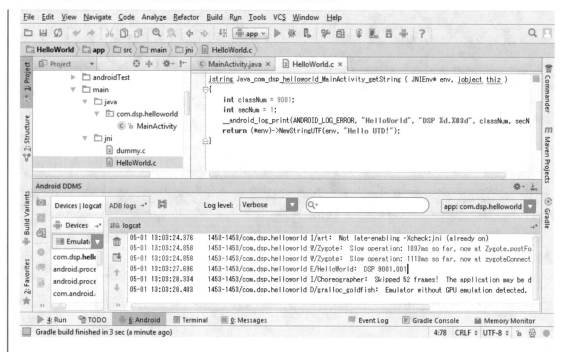

Figure 2.32: LogCat screen.

```
jstring Java_com_dsp_helloworld_MainActivity_getString ( JNIEnv* env,
    jobject thiz )  {
                    return (*env)->NewStringUTF(env, ''Hello UTD!'');
}
```

According to the JNI convention, the inputs to this method, i.e., JNIEnv* env and jobject thiz, are always required. Additional input variables may be added and the return type may be changed as noted below:

```
jfloat Java_com_dsp_helloworld_MainActivity_getArea ( JNIEnv* env,
    jobject thiz, jfloat radius)  {
                    return 3.14159f*radius*radius;
}
```

with the corresponding native method in Java declared as

```
public native float getArea(float radius);
```

CHAPTER 3

iOS Software Development Tools

This chapter covers the required steps for running C codes on iPhone smartphones. This chapter is the iOS version of Chapter 2 which detailed the Android app development. This time a "Hello World!" app in the iOS development environment is constructed.

C code segments are made available to the iOS Objective-C environment through the normal header files used in C. Objective-C allows C codes to run without the need for any external wrapper. For accessing inputs and outputs or sensor signals on iPhones, existing available iOS APIs in Objective-C are used.

The development environment consists of the Xcode IDE. This development environment allows writing C codes, compiling, and debugging on an iOS device simulator or on an actual iOS device. The Xcode IDE includes a built-in debugger that can be used to debug C codes line-by-line and also to observe values stored for different variables. Xcode is available as a free download on Mac machines through the Apple app store.

To develop iOS apps, the following are needed:

- an Apple Mac computer,

- enrollment in an Apple approved developer program, and

- an iOS device

Note that in the absence of an actual iOS device, the iOS simulator can still be used. Different iOS configurations can be selected from the scheme selector, which is located at the top left of the Xcode window.

3.1 APP DEVELOPMENT

1. Launch the Xcode IDE. You should be prompted with a splash screen, as shown in Figure 3.1. Select *Create a new Xcode project*.

   In case this screen is not displayed, you can use *File -> New -> Project*.

2. Select *iOS -> Application -> Single View Application* (see Figure 3.2). After clicking *Next*, the configuration of the project appears, as shown in Figure 3.3.

Figure 3.1: **Xcode.**

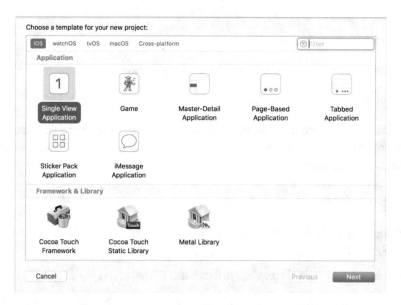

Figure 3.2: **Single View App.**

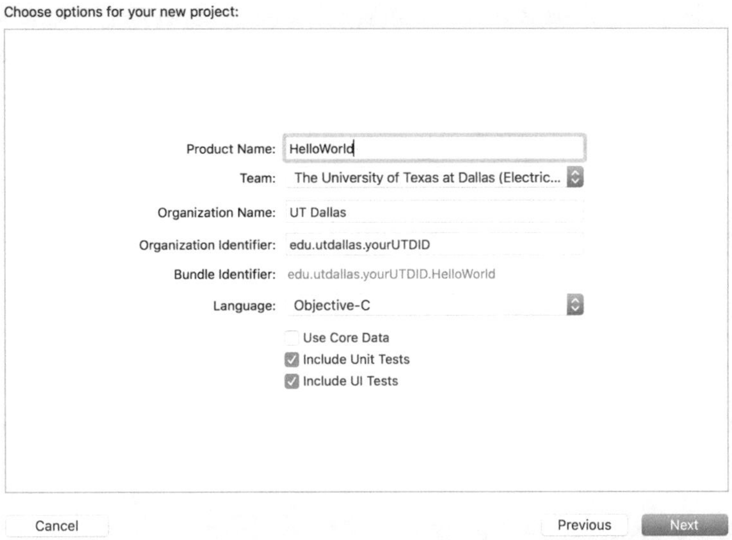

Choose options for your new project:

Product Name:	HelloWorld
Team:	The University of Texas at Dallas (Electric... ◇
Organization Name:	UT Dallas
Organization Identifier:	edu.utdallas.yourUTDID
Bundle Identifier:	edu.utdallas.yourUTDID.HelloWorld
Language:	Objective-C ◇

Use Core Data
☑ Include Unit Tests
☑ Include UI Tests

Cancel   Previous   Next

Figure 3.3: Configurations.

3. Enter *Product Name* as HelloWorld.

4. Enter an *Organization Name*.

5. Enter an *Organization Identifier*.

6. Set *Language* to Objective-C and *Devices* to iPhone.

7. Leave *Use Core Data* deselected.

8. Click *Next*. On the next page, remember to deselect *Create Git Repository on*.

9. Select the destination to store your project and select *Create*.

After clicking *Create*, the settings screen of the project gets shown. Here the features of the app can be altered, the devices supported by your project can be changed and also any additional frameworks or libraries to be utilized by your project can be added.

If getting a warning display "No signing identity found," this means you need to have your Apple Developer Account accepted for iOS app development. Also, your device must be certified for app development.

3.2   SETTING-UP APP ENVIRONMENT

The left column in the Xcode window is called the Navigator. Here one can select or organize different files and environment for a project.

- In the Navigator Pane, the Main.Storyboard entry is seen. This is used to design the layout of your app. Different UI elements in multiple views provided by the IDE can be used to design the interface of an app. However, this is done programmatically here.

- AppDelegate.h and AppDelegate.m are Objective-C files that can be used to handle events such as:

  - app termination
  - app entering background or foreground
  - app loading

  These files are not accessed here.

- The files ViewController.m and ViewController.h are used to define methods and properties specific to a particular view in the storyboard.

3.3   CREATING LAYOUT

In the file ViewController.m, the method called `viewDidLoad` is used to perform processes after the view is loaded successfully. This method is used here to initialize the UI elements.

In the interface section of ViewController, add the following two properties: a label and a button.

```
@interface ViewController ()
@property UILabel *label;
@property UIButton *button;
@end
```

Initialize the label and button and assign them to the view. This can be done by adding this code in the method `viewDidLoad`.

```
_label = [[UILabel alloc] initWithFrame:CGRectMake (10, 15, 300, 30)];

_label.text = @''Hello World!'';
[self.view addSubview:_label];
_button = [UIButton buttonWithType:UIButtonTypeRoundedRect];
_button.frame = CGRectMake(10, 50, 300, 30);
[_button setTitle:@''Button'' forState:UIControlStateNormal];
[self.view addSubview:_button];
[_button addTarget:self action:@selector(buttonPress:)
            forControlEvents:UIControlEventTouchUpInside];
```

An action is attached to the button `buttonPress` . This action will call a method which will be executed when the button is pressed. As the control event is `UIControlEventTouchUpInside` , the method will be executed when the user releases the button after pressing. The method is declared as follows:

```
(IBAction)buttonPress:(id)sender {
}
```

As of now, it is a blank method. A property will be assigned to it after specifying a C code that is to be used to perform a signal processing function.

The app can be run by pressing the *Play* button on the top left of the Xcode window. An actual target is not needed here and one may select a simulator; see Figure 3.4. For example, iPhone 6 can be selected as the simulator.

When the app is run, the label "Hello World!" can be seen and the button gets created. On clicking the button, nothing happens. This is because the method to handle the button press is empty.

It is not required to know the Objective-C syntax. The item to note here is that one can call a C function in Objective-C just by including a header. Objective-C is useful for handling iOS APIs for sound and video i/o, whereas signal processing codes can be written in C.

3.4   IMPLEMENTING C CODES

In this section, a C code is linked to *ViewController* using a header file.

- Right-click on the *HelloWorld* folder in your project navigator in the left column and select *New File*.

- Select *iOS -> Source -> C File*.

- Write the file name as Algorithm and select *Also create a header file*.

- After clicking *Next*, select the destination to store the files. Preferably store the files in the folder of your project.

- In the project navigator, you can view the two new added files. Select Algorithm.c.

- In Algorithm.c, enter the following C code:

```
const char *HelloWorld() {
      printf''Method Called'');
      return ''Hello UTD!'';
}
```

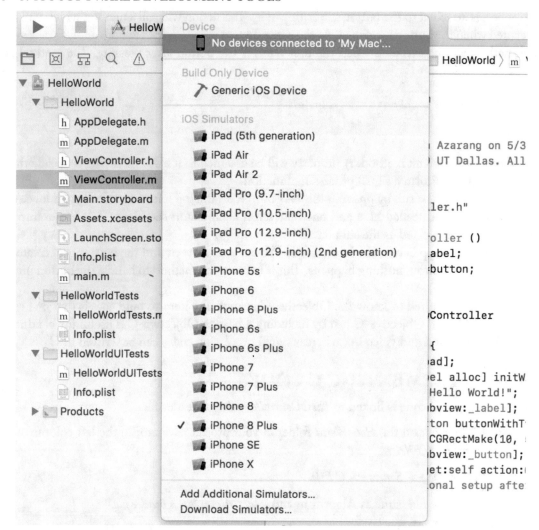

Figure 3.4: Simulator selection.

- The function `HelloWorld()` prints a string and returns a char pointer upon execution. Let us call this function on the button press action in the view controller and alter the label.

- To allow this function to be called in Objective-C, the function in the header file needs to be declared. For this purpose, in Algorithm.h, add the following line before `#endif`:

```
const char *HelloWorld();
```

- Now a C function is created, which is called and executed via Objective C, just by including the header file.

3.5   EXECUTING C CODES VIA OBJECTIVE-C

Now that a C code is written, it needs to be linked to the Objective-C app in order to be executed.

- In ViewController.m, just below `#import ''ViewController.h''`, add `#import ''Algorithm.h''`

- In the `buttonPress` method, include the following line:

```
_label.text = [NSString stringWithUTF8String:HelloWorld()];
```

This code line alters the text of the label in the program.

- Run the program in the simulator.

On pressing the button in the simulator, the following is observed.

1. The text of the label changes.

2. In the Xcode window, "Method Called" gets printed in the Debug Console at the bottom.

This shows that printing can be done from the C function to the debug console in Xcode. This feature is used for debugging purposes.

3.6   SWIFT PROGRAMMING LANGUAGE

Swift is a programming language developed by Apple as a successor to Objective-C. It provides a modern programming-language interface with a simple syntax, making it easier to read and maintain as compared to Objective-C. In Swift, pointers are not exposed by default to prevent memory leakage due to improper referencing or dereferencing of memory addresses. Swift allows less coding to be done to achieve the same objective as compared to Objective-C, for example:

- Creating a string and concatenating in Objective-C:

```
NSString *str = @''Hello'';

str = [str stringByAppendingString:@''World!''];
```

- Creating a string and concatenating in Swift.

```
var str = ''Hello''

str += ''World!''
```

Since declarations in Objective-C are required to be done as pointers, one needs to be careful with memory allocation and preventing memory leaks when using Objective-C. With Swift, these issues are dealt with in an easy manner. As noted above, via type inference, Swift can detect "Hello" as String, removing the need to explicitly state the datatype. As seen in the above Objective-C example, there are distinct compilation units similar to C with header files to pass variables and functions to other files. This is not required in Swift and the same is achieved by the declaration of only one .swift file and by declaring variables and functions only once.

Swift allows calling Objective-C and C functions via a bridging header. As a result, previously developed codes can be seamlessly called in a Swift program without the need for translation. This makes Swift suitable for developing app GUIs while running processing codes in C.

To develop a Swift project, the same Objective-C example is considered here. The following steps need to be taken to create the same example in Swift.

- Follow Section 3.1 until step 6 and set *Language* to Swift.

- To create the layout, open *ViewController.swift*. Note that there is only one file associated with the coding.

- Under the class definition and before the `viewDidLoad` function definition, initialize a UILabel and a UIButton via the following code. Note that one does not need to declare them as pointers:

```
class ViewController: UIViewController {
      var label: UILabel!
      var button: UIButton!
```

- Add the following code to the `viewDidLoad` function to programmatically create the UI similar to how it is done in Objective-C. Notice that there is no underscore before `label` and `button` . Also, memory does not need to be allocated for the labels, and functions can be chained simply by using '.' instead of '[]' used in Objectove-C. The Swift code appears below:

```
label = UILabel(frame:CGRect(x: 10, y: 15, width: 300, height: 30))
label.text = ''Hello World''
self.view.addSubview(label)

button = UIButton(frame: CGRect(x: 10, y: 50, width: 300,
                  height: 30))
button.setTitle(''Button'', for: .normal)
button.addTarget(self,  action:
  #selector(self.buttonClicked), for: .touchUpInside)
button.backgroundColor = .black
self.view.addSubview(button)
```

- Attach a function to the `buttonPress` function by adding the following function definition below the `viewDidLoad` function:

```
@objc func buttonPress() {
}
```

- The C code can be done similar to Section 3.4. When a prompt is raised to add a bridging header to the project, select "Yes".

- After the C coding, the Algorithm.h file can be included in the bridging header file. This allows C functions to be called in the Swift program.

- Update the `buttonPress` function by adding the following line of code to its function definition:

```
label.text = String(cString: HelloWorld())
```

- The program can then be built and run similar to the previously created example. On clicking the button on the GUI, the text changes to "Hello UTD!" and the Method Called is displayed in the Debug Console in Xcode.

More information about the Swift programming language are available at https://swift.org and the documentation for its uses in iPhone and MacOS are available at https://developer. apple.com/documentation/swift.

```
 9  #include "Algorithm.h"
10
11  const char *HelloWorld() {
12      printf("Method Called\n");
13      return "Hello UTD!";
14  }
15
```

Figure 3.5: Debug point.

L2   LAB 2:
iPHONE APP DEBUGGING

After getting a familiarity with the Xcode IDE by creating and modifying an iOS app project and running the app on an iPhone simulator, the following lab experiment can be done to debug C codes via the built-in Xcode debugger.

To obtain familiarity with the Xcode debugging tool, perform the following.

- Begin by acquiring the C code to be used for this lab.

- Open the folder containing the project.

- Double click on the file with the extension .xcodeproj.

- Navigate to the C code in the Project Navigator.

The app can be built by going to *Product -> Build*.

After the project is successfully built, debug points can be placed inside the C code. The debug points can be placed by clicking on the column next to the line to be debugged or by pressing CMD + \. A blue arrow appears (see Figure 3.5) that points toward the line to be debugged.

The Xcode debugger allows one to:

- pause the execution at a particular line of code,

- know the value of the variable at that particular instant of execution, and

- navigate from function call to function execution as it is executed.

The recent Xcode version at the time of this writing includes the LLDB debugger, which allows one to view data in an array with a pointer by typing the following command in the debug console after the debug point is encountered:

```
memory read -t [Data Type of the array] -c[Number of elements]
 '[Name of the array]'
```

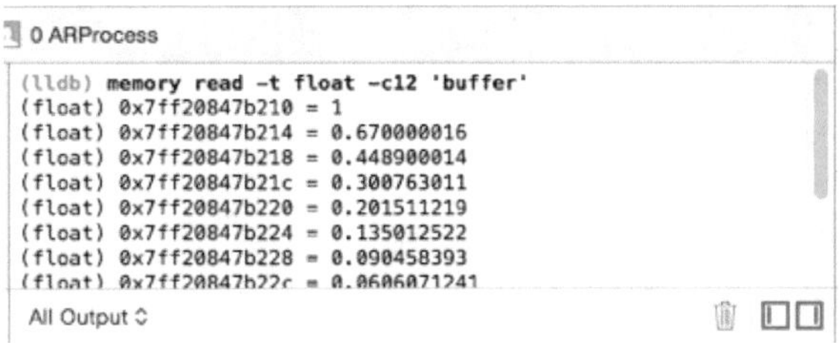

Figure 3.6: Output appearance.

For example, `memory read -t float -c12 'buffer'`.

When this command executes, the output appears as shown in Figure 3.6.

L2.1   LAB EXERCISE

Debug the broken C code named Lab2BrokenFilter using the Xcode debugger. This code is supposed to filter out the higher-frequency (3600 Hz) sinusoidal signal of a test signal made up of two sinusoids (the test signal previously loaded onto the device) and pass through its lower-frequency (500 Hz) sinusoidal signal. Indicate how both the syntax and logical errors in this code can be fixed.

Hints   When attempting to debug with Xcode, first get the code to a point where it will compile. Then, fix the logical errors in the code. The code will compile as is, but the processing functionality is commented out. When testing, the reference output found in lab2_testsignal.txt can be used to verify the result. This should match the output from the C code in your app.

CHAPTER 4

Analog-to-Digital Signal Conversion

The process of analog-to-digital signal conversion consists of converting a continuous time and amplitude signal into discrete time and amplitude values. Sampling and quantization constitute the steps needed to achieve analog-to-digital signal conversion. To minimize any loss of information that may occur as a result of this conversion, it is important to understand the underlying principles behind sampling and quantization.

4.1 SAMPLING

Sampling is the process of generating discrete time samples from an analog signal. First, it is helpful to see the relationship between analog and digital frequencies. Let us consider an analog sinusoidal signal $x(t) = A\cos(\omega t + \phi)$. Sampling this signal at $t = nT_s$, with the sampling time interval of T_s, generates the discrete time signal

$$x[n] = A\cos(\omega nT_s + \phi) = A\cos(\theta n + \phi), \qquad n = 0, 1, 2, \ldots, \tag{4.1}$$

where $\theta = \omega T_s = \frac{2\pi f}{f_s}$ denotes digital frequency with units expressed in radians (as compared to analog frequency ω with units expressed in radians/sec).

The difference between analog and digital frequencies is more evident by observing that the same discrete time signal is obtained for different continuous time signals if the product ωT_s remains the same. An example is shown in Figure 4.1. Likewise, different discrete time signals are obtained for the same analog or continuous time signal when the sampling frequency is changed. An example is shown in Figure 4.2. In other words, both the frequency of an analog signal and the sampling frequency define the frequency of the corresponding digital signal.

It helps to understand the constraints associated with the above sampling process by examining signals in frequency domain. The Fourier transform pairs in analog and digital domains are given by

$$\textit{Fourier transform pair for} \atop \textit{analog signals} \qquad \begin{cases} X(j\omega) = \int\limits_{-\infty}^{\infty} x(t)\,e^{-j\omega t}\,dt \\[2mm] x(t) = \frac{1}{2\pi} \int\limits_{-\infty}^{\infty} X(j\omega)\,e^{j\omega t}\,d\omega \end{cases} \tag{4.2}$$

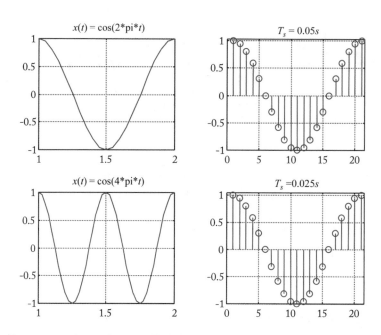

Figure 4.1: **Different sampling of two different analog signals leading to the same digital signal.**

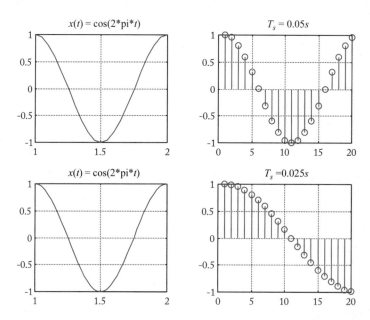

Figure 4.2: **Different sampling of the same analog signal leading to two different digital signals.**

Fourier transform pair for discrete signals

$$
\begin{cases}
X\left(e^{j\theta}\right) = \displaystyle\sum_{n=-\infty}^{\infty} x\left[n\right]e^{-jn\theta}, \quad \theta = \omega T_s \\
x\left[n\right] = \frac{1}{2\pi}\displaystyle\int_{-\pi}^{\pi} X\left(e^{j\theta}\right)e^{jn\theta}\,d\theta.
\end{cases}
\tag{4.3}
$$

As illustrated in Figure 4.3, when an analog signal with a maximum frequency of $f_{\max}$ (or bandwidth of W) is sampled at a rate of $T_s = \frac{1}{f_s}$, its corresponding frequency response is repeated every 2π radians, or f_s. In other words, Fourier transform in digital domain becomes a periodic version of Fourier transform in analog domain. That is why, for discrete signals, one is only interested in the frequency range $0-f_s/2$.

Therefore, in order to avoid any aliasing or distortion of the frequency content of the discrete signal, and hence to be able to recover or reconstruct the frequency content of the original analog signal, the sampling frequency must obey this rate $f_s \geq 2f_{\max}$. This is known as the Nyquist rate; that is, the sampling frequency should be at least twice the highest frequency in the signal. Normally, before any digital manipulation, a frontend anti-aliasing analog lowpass filter is used to limit the highest frequency of the analog signal.

Figure 4.4 shows the Fourier transform of a sampled sinusoid with a frequency of f_o. As can be seen, there is only one frequency component at f_o. The aliasing problem can be further illustrated by considering an under-sampled sinusoid as depicted in Figure 4.5. In this figure, a 1 kHz sinusoid is sampled at $f_s = 0.8$ kHz, which is less than the Nyquist rate. The dashed-line signal is a 200 Hz sinusoid passing through the same sample points. Thus, at this sampling frequency, the output of an A/D converter would be the same if either of the sinusoids were the input signal. On the other hand, over-sampling a signal provides a richer description than that of the same signal sampled at the Nyquist rate.

4.2   QUANTIZATION

An A/D converter has a finite number of bits (or resolution). As a result, continuous amplitude values get represented or approximated by discrete amplitude levels. The process of converting continuous into discrete amplitude levels is called quantization. This approximation leads to an error called quantization noise. The input/output characteristic of a 3-bit A/D converter is shown in Figure 4.6 to see how analog values get approximated by discrete levels.

The quantization interval depends on the number of quantization or resolution level, as illustrated in Figure 4.7. Clearly the amount of quantization noise generated by an A/D converter depends on the size of quantization interval. More quantization bits translate into a narrower quantization interval and hence into a lower amount of quantization noise.

To avoid saturation or out-of-range distortion, the input voltage must be between V_{ref-} and V_{ref+}. The full-scale (FS) signal V_{ref} is defined as

$$
V_{FS} = V_{ref} = V_{ref+} - V_{ref-}
\tag{4.4}
$$

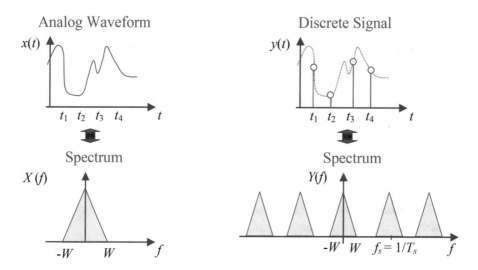

Figure 4.3: (a) Fourier transform of a continuous-time signal and (b) its discrete time version.

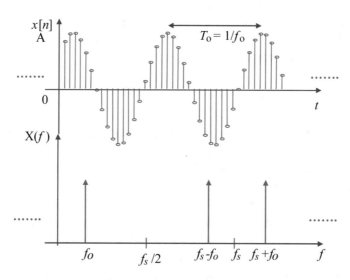

Figure 4.4: Fourier transform of a sampled sinusoidal signal.

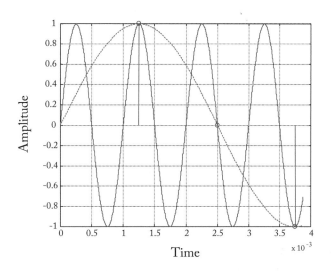

Figure 4.5: Ambiguity caused by aliasing.

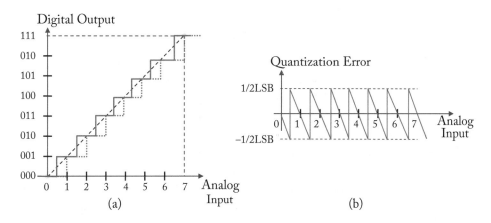

Figure 4.6: Characteristic of a 3-bit A/D converter: (a) input/output static transfer function and (b) additive quantization noise.

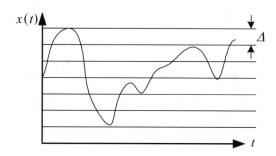

Figure 4.7: Quantization levels.

and one least significant bit (LSB) is given by

$$1LSB = \Delta = \frac{V_{ref}}{2^N}, \qquad (4.5)$$

where N is the number of bits of the A/D converter. Usually, it is assumed that quantization noise is signal independent and is uniformly distributed over -0.5 LSB and 0.5 LSB. Figure 4.8 shows the quantization noise of an analog signal quantized by a 3-bit A/D converter. It is seen that, although the histogram of the quantization noise is not exactly uniform, it is reasonable to consider the uniformity assumption.

L3   LAB 3: ANDROID AUDIO SIGNAL SAMPLING

This lab provides an understanding of the tools provided by the Android API for capturing audio signals and outputting processed audio signals. Android API documentation is available online at http://developer.android.com/reference/packages.html. The two relevant packages for this lab are android.media.AudioRecord for audio input and android.media.AudioTrack for audio output. As noted in the previous labs, the Android emulator does not support audio input. Additionally, the computation time on the emulator is not accurate or stable so an actual smartphone target is required in order to obtain proper computation times in the exercises.

This lab involves an example app demonstrating how to use the Android APIs supplied in Java and how to wrap C code segments so that they can be executed using the Java Native Interface (JNI). The example app records an audio signal from the smartphone microphone and applies a lowpass filter to the audio signal. An overview of the dataflow is shown in Figure 4.9.

Input samples can come from either a file or from the microphone input. These samples are stored in a Java WaveFrame object (a data wrapper class used for transferring of data). The BlockingQueue interface is used to transfer data from the input source to the processing code, and finally to the output destination (either file or speaker). Using the BlockingQueue interface is advantageous because it allows a small buffer for accumulating data, while at the same time

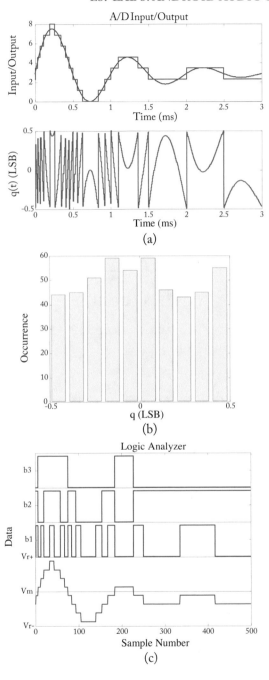

Figure 4.8: Quantization of an analog signal by a 3-bit A/D converter: (a) output signal and quantization error, (b) histogram of quantization error, and (c) bit stream.

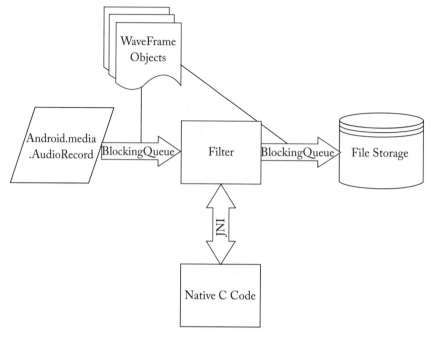

Figure 4.9: Dataflow.

functioning as a First-In-First-Out (FIFO) queue. The WaveFrame objects are helper objects which serve to store audio samples in an array. These objects are stored and re-used to reduce garbage collection in the Java VM.

L3.1   DEMO APPLICATION

For this lab, a prebuilt Android app APK is provided in the book software package. The app is built for compatibility with Android 3.5.3 ensuring compatibility with new versions of Android. The app is called *RealTime* and can be found in the Lab 3 folder in the software package. This application allows studying sampling rate and frame size and their effects on the maximum computation delay allowable in a real-time signal processing pipeline. It is required to use a real Android target for these experiments as the emulator timings do not reflect those of an actual target device. Additionally, the emulator does not allow testing of alternate sampling rates or audio recording. In order to install the application on your smartphone, it may be necessary to enable software installation from *Unknown sources*. The *Unknown sources* option can be found in the general *Settings* configuration on the smartphone. Look for the *Unknown sources* option found in the *Security* submenu.

The project source of the app is also provided for your reference. This app performs signal sampling using the smartphone microphone. It can also read in a previously sampled signal from

Wav or PCM files in the device storage. The input signal is accumulated into frames of audio data and then passed through JNI to C code for lowpass filtering. An artificial computational delay is added to the filtering code by the use of the `usleep` function within the C processing segment. The filtered signal is then passed back to Java and is saved to a file or played back through the target audio output. The app settings menu allows the adjustment of the sampling rate, frame size, and computational delay of the signal processing pipeline.

L3.2   APPLICATION CODE

The app manifest appears below.

```xml
<?xml version=''1.0'' encoding=''utf-8''?>
<manifest xmlns:android=''http://schemas.android.com/apk/res/android''
      package=''com.dsp.filter''
      android:versionCode=''1''
      android:versionName=''1.0''>
  <uses-sdk android:minSdkVersion=''11''
            android:targetSdkVersion=''11''/>
  <uses-permission
      android:name=''android.permission.WRITE_EXTERNAL_STORAGE''/>
  <uses-permission android:name=''android.permission.RECORD_AUDIO/>
  <application android:label=''@string/app_name''>
      <activity android:name=''.RealTime''
                android:label=''@string/app_name''
                android:screenOrientation=''portrait''
                android:configChanges=''keyboardHidden|orientation''>
          <intent-filter>
              <action android:name=''android.intent.action.MAIN'' />
              <category android:name=
                      ''android.intent.category.LAUNCHER''/>
          </intent-filter>
      </activity>
  </application>
</manifest>
```

An important aspect of the app manifest is enabling the correct permissions to perform file IO and record audio. Permissions are set with the `uses-permission` directive, where `RECORD_AUDIO` is defined for audio input and `WRITE_EXTERNAL_STORAGE` enables saving files to the device storage. A detailed list of the permissions can be found on the Android developer website at:

http://developer.android.com/reference/android/Manifest.permission.html

The code to be executed is referred to as an *Activity*—in this case named *.RealTime* in the manifest, which corresponds to the *RealTime.java* file. The manifest indicates namings which are necessary for hooking into the UI elements, defined in the layout. *RealTime.java* contains the code which hooks the UI elements and controls the app execution.

L3.3   RECORDING

The API that supplies audio recording capability is found in android.media.AudioRecord. A reference implementation of this API appears within the *WaveRecorder* class of the sampling code. To determine which sampling rates are supported by an Android device, the function `WaveRecorder.checkSamplingRate()` is called upon the app initialization. This function initializes the *AudioRecord* API with a preset list of sampling rates. If the sampling rate is not supported, an error will occur and that particular sampling rate will not be added to the list of supported sampling rates. Supported sampling rates as determined by the function `checkSamplingRate` are listed in the app settings menu.

To initialize the recorder, the size of the data buffer to be used by the recorder is first computed by calling the function `getMinBufferSize` as follows:

```
int BufferLength = AudioRecord.getMinBufferSize(FS, CHANNELS, FORMAT);
```

where FS specifies the desired sampling rate, CHANNELS is the channel configuration which can be either stereo or mono, and lastly FORMAT specifies 16-bit PCM or 8-bit PCM audio data. The return value of this function is the minimum size of the buffer in bytes. It is important to mention that the size of this buffer is dependent upon the Android target and will vary from target to target. To ensure that no audio data is lost, this value can be scaled up to ensure that there is enough overhead, but needs to be set to at least the size returned by `getMinBufferSize`. Scaling up the size of this buffer will also affect the latency of the recording; larger buffers will cause increased delay before the sampled audio is available for processing. The same is true for the *AudioTrack* API; increasing the output buffer size will increase the delay before the processed audio is outputted to the smartphone speaker. This BufferLength value is used to instantiate the AudioRecord object:

```
AudioRecord record = new AudioRecord(SOURCE, FS, CHANNELS, FORMAT,
    BufferLength);
```

The values used for FS, CHANNELS, and FORMAT should match those used to calculate the buffer length. There are several options for the SOURCE parameter, detailed at http://developer.android.com/reference/android/media/MediaRecorder.AudioSource.html. For most cases it should be specified as `AudioSource.CAMCORDER` as this ensures that the built-in filters for noise reduction and gain correction for voice calls are not active during the recording.

The audio recorder does not begin to accumulate data when it is instantiated and must be controlled. To begin collecting data, the function `recorder.startRecording()` is called. The recorder object is then polled to retrieve audio data by calling one of the read functions. If audio is not read from the recorder at a sufficient rate, the internal buffer will overflow. The read function must be supplied with a buffer to write data into, an initial offset, and a final offset. These functions are blocking—meaning that the program flow will not continue until the desired amount of audio data has been read into the supplied buffer. The recording loop is shown below:

```
loop:while(true){
    if(isRecording.get()) {
        out = recycleQueue.take();
        recorder.read(out.getAudio(), 0, Settings.stepSize);
        output.put(out);
    } else {
        output.put(Settings.STOP);
        break loop;
    }
}
```

In the reference code, `recycleQueue` is a BlockingQueue filled with a predetermined number of WaveFrame objects. The data are then inserted into a WaveFrame structure which are passed along to a queue for further processing in the Filter class. The `output.put` function attempts to insert the frame of audio data into the output queue if there is space available. If the output queue is full, it will wait to insert the data, effectively pausing the recording process. If the signal processing takes too long, the internal buffer of the recorder will fill up and cause an exception to be thrown. When audio recording is finished, the recorder object should be stopped and the resources released.

The WaveFrame class is mainly used as a helper to convert between data types and store diagnostic timing. BlockingQueue is configured to store up to ten frames of audio data, as defined in the *Settings.java* class by the `queueSize` variable. These WaveFrame objects will remain allocated while the program is running in order to prevent excessive memory allocations from taking place. Once they get processed and the end of the pipeline is reached, they are reinserted into the recycling queue for reuse.

L3.4   PROCESSING.JAVA

The code that is responsible for calling the native methods is found in the *Processing.java* class. When the app runs, the recorder inserts audio data into the input queue. The *Processing* class polls that queue to retrieve a frame of audio data, processes it, and inserts the output data into the output queue. The loop that accomplishes these is indicated below:

```
loop:while(true) {
WaveFrame currentFrame = null;
currentFrame = input.take();
if(currentFrame == Settings.STOP){
        output.put(currentFrame);
        break loop;
}

        process(currentFrame.getAudio());
        getOutput(currentFrame.getAudio(), Settings.output);
output.put(currentFrame);
}
```

`process()` takes the short array input corresponding to the input signal and processes it using the native code methods. After the processing method is called, the samples corresponding to the filtered output will be stored temporarily in a memory location allocated by the native code. To retrieve the filtered signal, the getOutput() method is used to overwrite the contents of the WaveFrame audio buffer with the desired output as selected by an integer-based switch.

L3.5   JNI NATIVE C CODE

The above explanation on Java indicated how to write a wrapper around C codes. There are three basic steps that need to be followed when writing C codes using JNI.

1. Native code definitions—within one of the Java classes, it is required to have code declarations which match the signature (inputs and outputs match types) of the functions implemented in the native code. In the example app, these definitions appear in the Filters class. Another example is shown below:

```
package <tld>.<your_domain>.<your_package>;
public class <YourClass> {
        public static native float[]
                <yourMethod>(float[] in, float a, int b);
}
```

2. Method naming within the native code should follow a set pattern and handle JNI variables as indicated below. Note that the input and output types shown in the above

`yourMethod` declaration correspond to the input and output types in the declaration noted below with the addition of some JNI fields.

```
jfloatArray
Java_<tld>_<your_domain>_<your_package>_<YourClass>_<yourMethod>
(JNIEnv *env, jobject thiz, jfloatArray in, jfloat a, jint a)
{ /*do stuff*/ }
```

3. Loading native libraries into memory when the program runs. The following code should be placed in the class file responsible for defining the Android activity. In the example app, this can be found in the GUI class:

```
static {
        System.loadLibrary(''yourlibrary'');
}
```

`System.loadLibrary` links the native method declaration in the Java class with the processing code declared in the C file. There are alternative methods of declaring native functions and performing the linking which will be covered in a later chapter.

Depending on the input data types in the methods and the expected return data types, the JNI method signature will change. If any of these steps are not correctly done, Java will not be able to find the native methods and an exception will be thrown when the app attempts to call the native method.

L3.6   SUPERPOWERED SDK

For real-time low-latency througputs on Android devices, the Superpowered utility can be used. This utility is an audio API which is developed for mobile devices. On Android devices, Superpowered uses OpenSL to allow processing of audio data in a low-latency manner. More information on these APIs are available at the following links:

- Superpowered: http://superpowered.com

- OpenSL ES: https://www.khronos.org/opensles/

In what follows, a simple audio I/O path is implemented via Superpowered and it is shown how to implement a low-latency Android app.

- Go to http://superpowered.com and download the SuperpoweredSDK. Store it in *C:\Android*.

Figure 4.10: FrequencyDomain example.

- Open Android Studio. In the splash screen, select *Open an existing Android Studio Project*.

- Navigate to the *SuperpoweredSDK* folder. In the *Android* subfolder, open the *Frequen-cyDomain* example in Android Studio, as shown in Figure 4.10.

- Open *MainActivity.java*. In that, the JNI native function `FrequencyDomain` can be seen, which is the C++ function that calls the function to start audio I/O with the supplied sampling rate and input buffer size. As it gets called inside the `onCreate` method, the audio path is created as soon as the app is loaded.

- Next, go to the FrequencyDomain.cpp file inside *cpp* in the project navigator. In this file, two methods can be seen:

  - `FrequencyDomain` : This is the JNI method which is called from *MainActivity.java*. This function is used to create a Superpowered Audio I/O session.

  - `audioProcessing` : This is the callback linked with the Audio I/O session which is repeatedly called when the input data are available for processing.

- To build a simple unprocessed audio i/o path, let us add a simple C code to process the incoming audio. Right-click on the CPP folder and go to *New > C/C++ Source File*. Enter the name of the file as *FIR*, set the type as *.c*, and click on *create an associated header*. The files may not appear in the Project Navigator. Add the following codes to the *CMakeLists.txt* file.

  – Add the following line of code:

  ```
  file(GLOB C_FILES ''*.c'')
  ```

  – In the add_library section, add the following line of code:

  ```
  ${C_FILES}
  ```

  After this is done, upon syncing the gradle, the two files should appear in the *jni* folder.

- In the *FIR.c* file, add the following code:

```c
void FIR(float* input, float* output, int nSamples) {
    int i = 0;

    static float endSamples[2] = {0,0};

    for (i = nSamples - 1; i > 1; i--) {
        output[i] = (input[i] + input[i - 1] + input[i - 2])/3;
    }

    output[1] = (input[1] + input[0] + endSamples[1])/3;
    output[0] = (input[0] + endSamples[1] + endSamples[0])/3;

    endSamples[1] = input[nSamples - 1];
    endSamples[0] = input[nSamples - 2];

}
```

In the *FIR.h* file, add the following code:

```
void FIR(float* input, float* output, int nSamples);
```

- After the algorithm is coded in C, it can be called in the main file. In *FrequencyDomain.cpp*, replace the existing code with the following code:

```
#include <jni.h>
#include <Superpowered.h>
#include <SuperpoweredFrequencyDomain.h>
#include <OpenSource/SuperpoweredAndroidAudioIO.h>
#include <SuperpoweredSimple.h>
#include <SuperpoweredCPU.h>
#include <SLES/OpenSLES.h>
#include <SLES/OpenSLES_AndroidConfiguration.h>
#include <cstring>
#include <cstdlib>

static SuperpoweredAndroidAudioIO *audioIO;

// To Call C functions
extern ''C'' {
#include ''FIR.c''
}
// Globally declared audio buffers
static float *inputBufferFloat, *leftInputBuffer,
    *rightInputBuffer, *leftOutputBuffer, *rightOutputBuffer;
// This is called periodically by the media server.
static bool audioProcessing(void * __unused clientdata,
    short int *audioInputOutput, int numberOfSamples,
    int __unused samplerate)
{
    Superpowered::ShortIntToFloat(audioInputOutput,
                inputBufferFloat,  numberOfSamples, 2);
    Superpowered::DeInterleave(inputBufferFloat, leftInputBuffer,
                rightInputBuffer, numberOfSamples);
    FIR(leftInputBuffer, leftOutputBuffer, numberOfSamples);
    FIR(rightInputBuffer, rightOutputBuffer, numberOfSamples);
    Superpowered::FloatToShortIntInterleave(leftOutputBuffer,
                rightOutputBuffer, audioInputOutput,
```

```
                          numberOfSamples);
    return true;
}

// FrequencyDomain - Initialize buffers and setup frequency
//                   domain processing.
extern ''C'' JNIEXPORT void
Java_com_superpowered_frequencydomain_MainActivity_FrequencyDomain
(JNIEnv * __unused env, jobject __unused obj, jint samplerate,
 jint buffersize) {
    Superpowered::Initialize(
            ''ExampleLicenseKey-WillExpire-OnNextUpdate'',
            false,
            true,
            false,
            false,
            false,
            false,
            false
    );

    inputBufferFloat = (float *)malloc(buffersize *
                               sizeof(float) * 2 + 128);
    leftInputBuffer = (float *)malloc(buffersize *
                               sizeof(float) + 128);
    rightInputBuffer = (float *)malloc(buffersize *
                               sizeof(float) + 128);
    leftOutputBuffer = (float *)malloc(buffersize *
                               sizeof(float) + 128);
    rightOutputBuffer = (float *)malloc(buffersize *
                                sizeof(float) + 128);
    Superpowered::CPU::setSustainedPerformanceMode(true);
    audioIO = new SuperpoweredAndroidAudioIO (
            samplerate,      // native sampe rate
            buffersize,      // native buffer size
            true,            // enableInput
            true,            // enableOutput
            audioProcessing, // process callback function
```

```
                    NULL,               // clientData
                    -1,                 // inputStreamType (-1 = default)
                    SL_ANDROID_STREAM_MEDIA
                                        // outputStreamType (-1 = default)
        );
    // Start audio input/output.
    }


    // Cleanup - Stop audio processing and free resources.
    extern ''C'' JNIEXPORT void
    Java_com_superpowered_frequencydomain_MainActivity_Cleanup(
      JNIEnv * __unused env, jobject __unused obj) {
        delete audioIO;
        free(inputBufferFloat);
        free(leftInputBuffer);
        free(rightInputBuffer);
        free(leftOutputBuffer);
        free(rightOutputBuffer);
    }
```

In the above code, the `SuperpoweredAudioIO` session is initialized and then in `audioProcessing` the audio data are processed. The audio received is interleaved and it is deinterleaved before processing. After the processing is complete, the audio is interleaved again and stored back in the original buffer `audioInputOutput`. When true is returned in the method, it sends the buffer to the speaker.

- Run the app and listen to the audio path. Similar to *FIR.c*, one can add other audio processing algorithms to process audio data.

L3.7   MULTI-THREADING

The audio processing in Android can be run on a separate thread allowing the audio to be processed uninterrupted. Any computation that takes more time to run than the minimum allotted frame time can cause frames to get skipped. The use of Java Virtual Machine (JVM) allows having multiple threads in apps. This means that one canoffload processes that do not need to be run at frame rate to concurrent threads and thus allowing the audio to run without interruptions. Android has a thread class in Java (https://developer.android.com/reference/java/lang/Thread) that allows creating and monitoring threads for concurrently processing data.

As a simple example, let us create a multi-threading app that counts the number of frames processed by the Superpowered app in the previous section.

- First, create a hook to the TextView on the main page of the GUI via assigning an id to it. This can be done by going to *activity_main.xml* and adding the following code line to the TextView layout:

```
android:id=''@+id/mainactivityTextView''
```

- After providing the TextView with an id, instantiate it in the *MainActivity.java* file by adding the following code line in the `MainActivity` class:

```
TextView mainActivityTv;
```

and linking it with the GUI element by adding the following code line in the `onCreate` function:

```
mainActivityTv = (TextView)
            findViewById(R.id.mainActivityTextView);
```

- To update the TextView with the number of frames that are processed, a JNI function needs to be created. In the *FrequencyDomain.cpp*, create this static global variable

```
static int frameCount = 0;
```

and update it in the audioProcessing function this way:

```
frameCount++;
```

The frameCount variable can be accessed via TextView by creating the following JNI function:

```
extern ''C''

JNIEXPORT jint JNICALL

Java_com_superpowered_frequencydomain_MainActivity_GetFrameCount
(JNIEnv *env, jobject instance) {
```

```
        return frameCount;
    }
```

- After the JNI function is created, declare it as a native function in the *MainActivity.java* file in the `MainActivity` class as noted below:

```
private native int GetFrameCount();
```

- Then, declare the thread for updating TextView every second and outputting the number of frames processed as follows:

```
Thread thread = new Thread(new Runnable() {
    @Override
    public void run() {
        while(true) {
            try{
                sleep(1000);
                runOnUiThread(new Runnable() {
                    @Override
                    public void run() {
                        mainActivityTv.setText(
                        ''Number of frames processed:'' +
                        GetFrameCount() + '' frames'');
                    }
                });
            } catch (Exception e) {
                e.printStackTrace();
            }
        }
    }
});
```

In this code, the Thread starts a timer that runs every second using the sleep function and updates TextView in the GUI in real-time.

- After the thread is created, it can be started by declaring its start method in the last line of the `onCreate` function noted below:

```
    thread.start();
```

After doing the above, run the project. On the GUI, the total number of frames is displayed. This shows that the GUI can get updated while running the audio processing on a separate or different thread.

L3.8   MULTI-RATE SIGNAL PROCESSING

Latency is the delay between the time an audio signal is captured and the time it is outputted. The time required to collect samples in a frame and the time to process these samples affect the amount of latency. Apps with low audio latency is desired for many audio applications. Excessive latency causes mismatches between the utterance of words and hearing them. In many audio application, it is desired to keep latency low say below 15 ms.

For Android smartphones, the audio latency varies from device to device due to different i/o hardware used. The ADC (Analog to Digital Converter) hardware in modern smartphones provides the lowest audio latency at 48 kHz sampling rate. However, since many audio signal processing is done at 16 kHz sampling rate or lower, one needs to synchronize the audio i/o hardware and the audio processing software in order to have the lowest latency that a typical smartphone offers. This synchronization can be achieved by multi-rate signal processing as illustrated in Figure 4.11.

For low-latency operation, the audio needs to be captured from the smartphone microphone at 48 kHz sampling rate and by using the frame size that is recommended by the manufacturer of a particular Android smartphone. Circular buffers are used to achieve the recommended frame size for input and output. For audio processing modules, the captured frames are downsampled to 16 kHz to match with the rate at which audio processing modules operate. After audio frames are processed, they are then up-sampled and outputted through the smartphone speakers at 48 kHz. This multi-rate approach is discussed in more detail in [1] whose app codes are publicly available. A comprehensive list of Android smartphones and their preferred frame sizes are listed at https://superpowered.com/latency.

L3.9   LAB EXERCISES

1. How much memory is required to buffer a 1-s length recording of audio signal samples? State your assumptions when determining this number.

2. Experiment with various frame sizes and computation delays to find the maximum acceptable computation delay for a given frame size. Explain the situations when real-time processing breaks down.

Hints   Here are the steps that need to be taken for running the prebuilt app code on your Android smartphone target.

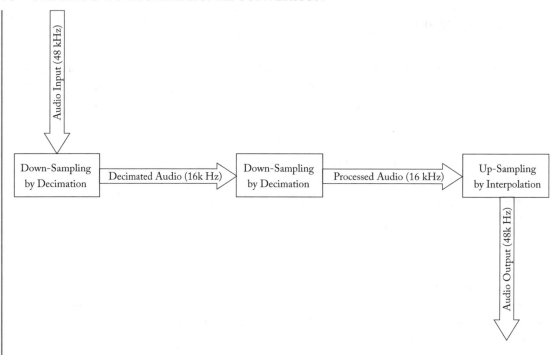

Figure 4.11: Multi-rate processing.

- Enable application installation from unknown sources. On Android 4.3, this option is in the *Settings > More > Security* section of the *Settings* menu. This option may also be in *Settings > Applications* on older Android versions.

- Put the application APK on the smartphone and open the APK to install the app.

You should then be able to run the app. You can record your own audio samples by selecting a sampling rate from the app *Settings* menu. Set the *Debugging Level* to *Wave* for the audio to be saved. Test the case when recording audio and the processing takes too long, and then the case when reading from a file and the processing takes too long.

L4   LAB 4:
iPHONE AUDIO SIGNAL SAMPLING

This lab is the iOS version of Lab 3 for capturing audio signals and outputting processed audio signals on an iPhone smartphone target. The iOS API documentation is available online at https://developer.apple.com/documentation/. The relevant framework for this lab is AudioToolbox. As noted previously, the iPhone simulator does not support audio input. Additionally, the

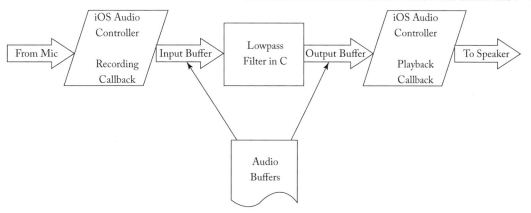

Figure 4.12: Audio signal dataflow.

computation time on the simulator is not accurate or stable, thus an actual smartphone target is required in order to obtain actual computation times in the exercises.

The structure of the previous debugging lab is re-used here. This lab involves an example app demonstrating how to use the iOS APIs supplied in Objective-C and how to properly link C code segments so that they can be executed using just a header file. The example app records an audio signal from the smartphone microphone and applies a lowpass filter to the audio signal. An overview of the dataflow is shown in Figure 4.12.

The input samples can come from either a file or from the microphone input. They get stored as short arrays in an AudioBuffer. The i/o buffers are used to transfer data from the input source to the processing code, and finally to the speaker as output. The use of callbacks allows one to input data from either the microphone or a file, store them in a software buffer, process and then output to the speaker.

L4.1   APP SOURCE CODE

The project source of this app allows one to study sampling rate and frame size and their effects on the maximum computation delay allowable in a real-time signal processing pipeline. It is required to use a real iOS device for this app, as the simulator timings do not reflect those of an actual target device.

This app performs signal sampling using the microphone of an iPhone smartphone. It can also read in a CAF or PCM file format into the device storage. The input signal is accumulated into frames of audio data and then passed to the C code for lowpass filtering. An artificial computation delay is added to the filtering code using the function *usleep* within the C code segment. The app settings menu allows adjusting the sampling rate, frame size, and delay.

L4.2   APP CODE DISCUSSION

iOS does not require any permissions to be added to a manifest. It automatically asks the user for permission to access the microphone. As the audio file provided for testing with this app is included in the main bundle, no special permissions are required to access it.

The code that is executed is called from the main.m file. It returns the app main in an autoreleasepool, i.e., it handles the auto releasing of memory variables.

```
#import <UIKit/UIKit.h>
#import ''AppDelegate.h''
#import ''IosAudioController.h''

int main(int argc, char * argv[]) {
    iosAudio = [[IosAudioController alloc] init];
    @autoreleasepool {
        return UIApplicationMain(argc, argv, nil,
            NSStringFromClass([AppDelegate class]));
    }
}
```

It calls on AppDelegate first to handle the app.

L4.3   RECORDING

The API that supplies the audio recording capability is part of the AudioToolbox. It can be included within a file using the instruction #import. To specify the audio format to be used by the app, an instance of AudioStreamBasicDescription needs to be created. This specifies the audio data to be processed.

```
AudioStreamBasicDescription audioFormat;
audioFormat.mSampleRate       = 44100.0;
audioFormat.mFormatID         = kAudioFormatLinearPCM;
audioFormat.mFormatFlags      =( kAudioFormatFlagIsSignedInteger |
                                 kAudioFormatFlagIsPacked);
audioFormat.mFramesPerPacket  = 1;
audioFormat.mChannelsPerFrame = 1;
audioFormat.mBitsPerChannel   = 16;
audioFormat.mBytesPerPacket   = 2;
audioFormat.mBytesPerFrame    = 2;
```

The above specifications indicate to the device that the audio data being handled has a sampling rate of 44.1 KHz, is linear PCM and is packed in the form of 16-bit integers. The audio is mono with only one frame per packet. The frame size is not determined here because the hardware determines the size of the frame at runtime.

This description is used to initialize the Audio Unit that will handle the audio i/o. The audio data is handled by a separate C function called a callback which is shown below:

```c
static OSStatus recordingCallback(void *inRefCon,
                                  AudioUnitRenderActionFlags *ioActionFlags,
                                  const AudioTimeStamp *inTimeStamp,
                                  UInt32 inBusNumber,
                                  UInt32 inNumberFrames,
                                  AudioBufferList *ioData) {


    // Because of the way audio format (setup below) is chosen:
    // only need 1 buffer, since it is mono
    // samples are 16 bits = 2 bytes
    // 1 frame includes only 1 sample

    AudioBuffer buffer;

    buffer.mNumberChannels = 1;
    buffer.mDataByteSize = inNumberFrames * 2;
    buffer.mData = malloc( inNumberFrames * 2 );

    // Put buffer in a AudioBufferList
    AudioBufferList bufferList;
    bufferList.mNumberBuffers = 1;
    bufferList.mBuffers[0] = buffer;

    // Then:
    // Obtain recorded samples

    if(getMic()) {
        OSStatus status;
        status = AudioUnitRender([iosAudio audioUnit],
                                 ioActionFlags,
                                 inTimeStamp,
                                 inBusNumber,
```

```
                                    inNumberFrames,
                                    &bufferList);
        checkStatus(status);

        // Now, we have the samples we just read sitting
        // in buffers in bufferList. Process the new data
        TPCircularBufferProduceBytes(inBuffer,
                    (void*)bufferList.mBuffers[0].mData,
                    bufferList.mBuffers[0].mDataByteSize);

        if(inBuffer->fillCount >= getFrameSize()*sizeof(short)) {
            [iosAudio processStream];
        }
    } else {
        UInt32 frameCount = getFrameSize();
        OSStatus err = ExtAudioFileRead(fileRef, &frameCount,
                                        &bufferList);
        CheckError(err,''File Read'');
        if(frameCount > (0) {

            AudioBuffer audioBuffer = bufferList.mBuffers[0];

            TPCircularBufferProduceBytes(inBuffer,
                    audioBuffer.mData,
                    audioBuffer.mDataByteSize);

            if (inBuffer->fillCount >=
                getFrameSize()*sizeof(short)) {
                [iosAudio processStream];
            }
        } else {
            getTime();
            [iosAudio stop];
            enableButtons();
        }
    }
}

// release the malloc'ed data in the buffer created earlier
```

```
        free(bufferList.mBuffers[0].mData);
        return noErr;
}
```

As the device collects data, the recording callback is called. Audio samples are collected and stored in a software buffer. The function processStream is used to process the audio samples in the software buffer.

```
- (void) processStream {

        //Frame Size
        UInt32 frameSize = getFrameSize() * sizeof(short);
        int32_t availableBytes;

        //Initialize Temporary buffers for processing
        short *tail = TPCircularBufferTail(inBuffer, &availableBytes);

        if (availableBytes > frameSize){
                short *buffer = malloc(frameSize),
                        *output = malloc(frameSize);
                memcpy(buffer, tail, frameSize);
                TPCircularBufferConsume(inBuffer, frameSize);
                FIRFilter(buffer, frameSize/sizeof(short), output,
                                getDelay());
                TPCircularBufferProduceBytes(outBuffer, output, frameSize);

                free(output);
                free(buffer);

                duration+=clock() - startTime;
                count++;
                startTime = clock();

                if(count > getValue()/getFrameSize()){
                        printTime((double)duration/((double)count *
                                                CLOCKS_PER_SEC));
                        duration = 0;
                        count = 0;
                }
```

```
        }
}
```

In this function, the capacity of the software buffer is checked with the frame size specified by the user. If the data in the software buffer is larger than the frame size, the input audio data is copied to the buffer and an empty buffer of the same size called output is created. Then, the C function is called to process the data provided in the buffer and puts it out. Finally, the data provided by the buffer is copied to the software buffer and it is outputted to the speaker.

L4.4   NATIVE C CODE

The C code to process the data can be linked to the file IosAudioController.m by including the corresponding header file. The header file includes all the global variables and functions to be linked outside their scope.

The steps required to link the C code are as follows.

- The function to be linked is coded in the .c file and the corresponding header file is linked with the same.

```
//FIRFilter.c
#include ''FIRFilter.h''

void FIRFilter(short* buffer, int frameSize, short* output,
               int delay)
{
        //Function Definition
}
```

- Function declaration is provided in the header file through which it can be linked to other files importing it.

```
//FIRFilter.h

#include <stdio.h>

void FIRFilter(short* buffer, int frameSize, short* output,
    int delay);
```

Once the function is declared in the header file, it can be accessed anywhere in the app, simply by importing the same header file.

L4.5   MULTI-THREADING

Similar to Android, iOS also requires all audio processing to occur within the minimum allotted frame time. iOS uses the Dispatch (https://developer.apple.com/documentation/DISPATCH) framework to allow processes to run concurrently on multicore processors. Dispatch allows running processes concurrently synchronously or asynchronously. This is done by offloading processes that are not required to run at frame time, allowing audio processing to be run uninterrupted and without any frame skipping.

As an example, the function which updates the iPhone GUI is stated below.

- Navigate to the GlobalVariable.m file in the Xcode project used above.

- The function printTime has the code that offloads the GUI update code to the main thread.

```
void printTime(double duration){
    dispatch_async(dispatch_get_main_queue(), ^(void){
        textView = (__bridge UITextView *)(view);
        NSString *status = [NSString stringWithFormat:
                @''Average Frame
Processing Time : %f milliseconds\n'', (double)duration*1e3];
        textView.text =
            [textView.text stringByAppendingString:status];
        [textView
            scrollRangeToVisible:NSMakeRange([[textView text]
            length], 0)];
    });
}
```

In this code, the dispatch_async function runs the code inside it on the main thread. This allows the audio processing code to run while the GUI is updated at the same time.

L4.6   MULTI-RATE SIGNAL PROCESSING

Similar to Android smarphones, iPhones also suffer from audio latency. The latency can be minimized by running at a sampling rate of 48 kHz similar to Android smartphones. The optimum frame size for all iOS devices is 64 samples and the round-trip latency ranges from

9–11 ms. The same multi-rate signal processing approach in Section L3.8 can be used here to synchronize the audio i/o and processing modules while generating a low-latency outcome.

L4.7   LAB EXERCISES

1. How much memory is required to buffer a 1-s length recording of audio signal samples?

2. Experiment with various frame sizes and computation delays to find the maximum allowable computation delay for a given frame size. Explain the situations when real-time processing breaks down.

Hints   Here are the steps that need to be taken for running the prebuilt app code on your iOS smartphone target:

- Select your device in the scheme window of the Xcode.

- Run the project. This will install the app.

You should then be able to run the app. You can record your own audio samples by selecting a sampling rate from the app Settings menu. Test the case when recording audio and the processing takes too long, and then the case when reading from a file and the processing takes too long.

4.5   REFERENCES

[1] A. Sehgal and N. Kehtarnavaz, Utilization of two microphones for real-time low-latency audio smartphone apps, *Proc. of IEEE International Conference on Consumer Electronics (ICCE)*, Las Vegas, NV, January 2018. DOI: 10.1109/icce.2018.8326213. 75

CHAPTER 5

Fixed-Point vs. Floating-Point

One important feature that distinguishes different processors is whether their CPUs perform fixed-point or floating-point arithmetic. In a fixed-point processor, numbers are represented and manipulated in integer format. In a floating-point processor, in addition to integer arithmetic, floating-point arithmetic can be handled. This means that numbers are represented by the combination of a mantissa (or a fractional part) and an exponent part, and the CPU possesses the necessary hardware for manipulating both of these parts. As a result, in general, floating-point operations involve more logic elements (larger ALU) and more cycles (more time) to manipulate floating-point values.

In a fixed-point processor, one needs to be concerned with the dynamic range of numbers, since a much narrower range of numbers can be represented in integer format as compared to floating-point format. For most applications, such a concern can be virtually ignored when using a floating-point processor. Consequently, fixed-point processors usually demand more coding effort than do floating-point processors.

5.1 Q-FORMAT NUMBER REPRESENTATION

The decimal value of a 2's-complement number $B = b_{N-1}b_{N-2} \ldots b_1 b_0$, $b_i \in \{0, 1\}$, is given by

$$D(B) = -b_{N-1}2^{N-1} + b_{N-2}2^{N-2} + \cdots + b_1 2^1 + b_0 2^0. \qquad (5.1)$$

The 2's-complement representation allows a processor to perform integer addition and subtraction by using the same hardware. When using unsigned integer representation, the sign bit is treated as an extra bit. Only positive numbers get represented this way.

There is a limitation to the dynamic range of the foregoing integer representation scheme. For example, in a 16-bit system it is not possible to represent numbers larger than $+2^{15} - 1 = 32,767$ or smaller than $-2^{15} = 32,768$. To cope with this limitation, numbers are normalized between -1 and 1. In other words, they are represented as fractions. This normalization is achieved by the programmer moving the implied or imaginary binary point (note that there is no physical memory allocated to this point), as indicated in Figure 5.1. This way, the fractional value is given by

$$F(B) = -b_{N-1}2^0 + b_{N-2}2^{-1} + \cdots + b_1 2^{-(N-2)} + b_0 2^{-(N-1)}. \qquad (5.2)$$

This representation scheme is referred to as Q-format or fractional representation. The programmer needs to keep track of the implied binary point when manipulating Q-format numbers. For instance, let us consider two Q15 format numbers and a 16-bit wide memory. Each

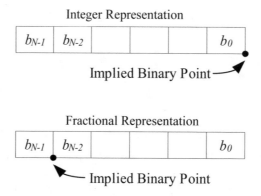

Figure 5.1: Integer vs. fractional representation.

number consists of 1 sign bit plus 15 fractional bits. When these numbers are multiplied, a Q30 format number is obtained (the product of two fractions is still a fraction), with bit 31 being the sign bit and bit 32 another sign bit (called extended sign bit). If not enough bits are available to store all 32 bits, and only 16 bits can be stored, it makes sense to store the most significant bits. This translates into storing the upper portion of the 32-bit product register, minus the extended sign bit, by doing a 1-bit left shift followed by a 16-bit right shift. In this manner, the product would be stored in Q15 format (see Figure 5.2). Notation for Q-format numbers is $QM \cdot N$ where M represents the number of bits corresponding to the whole-number part and N the number of bits corresponding to the fractional-number part.

Based on 2's-complement representation, a dynamic range of $-(2^{N-1}) \leq D(B) < 2^{N-1} - 1$ can be achieved, where N denotes the number of bits. For an easy illustration, let us consider a 4-bit system where the most negative number is -8 and the most positive number 7. The decimal representations of the numbers are shown in Figure 5.3. Notice how the numbers change from most positive to most negative with the sign bit. Since only the integer numbers falling within the limits -8 and 7 can be represented, it is easy to see that any multiplication or addition resulting in a number larger than 7 or smaller than -8 will cause overflow. For example, when 6 is multiplied by 2, the number 12 is obtained. Hence, the result is greater than the representation limits and will be wrapped around the circle to 1100, which is -4.

Q-format representation addresses this problem by normalizing the dynamic range between -1 and 1. Any resulting multiplication falls within the limits of this dynamic range. Using Q-format representation, the dynamic range is divided into 2^N sections, where $2^{-(N-1)}$ is the size of a section. The most negative number is always -1 and the most positive number is $1 - 2^{-(N-1)}$.

The following example helps one to see the difference in the two representation schemes. As shown in Figure 5.4, the multiplication of 0110 by 1110 in binary is the equivalent of multiplying 6 by -2 in decimal, giving an outcome of -12, a number exceeding the dynamic range

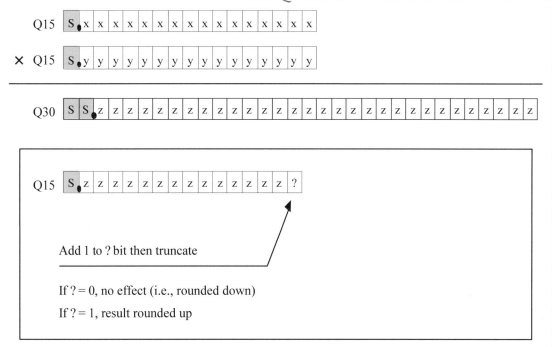

Figure 5.2: Multiplying and storing Q-15 numbers.

of the 4-bit system. Based on the Q3 representation, these numbers correspond to 0.75 and −0.25, respectively. The result is −0.1875, which falls within the fractional range. Notice that the hardware generates the same 1's and 0's; what is different is the interpretation of the bits.

When multiplying QN numbers, it should be remembered that the result will consist of $2N$ fractional bits, one sign bit, and one or more extended sign bits. Based on the data type used, the result has to be shifted accordingly. If two Q15 numbers are multiplied, the result will be 32 bits wide, with the MSB being the extended sign bit followed by the sign bit. The imaginary decimal point will be after the 30th bit. After discarding the extended sign bit with a 1-bit left shift, a right shift of 16 is required to store the result in a 16-bit memory location as a Q15 number. It should be realized that some precision is lost, of course, as a result of discarding the smaller fractional bits. Since only 16 bits can be stored, the shifting allows one to retain the higher precision fractional bits. If a 32-bit storage capability is available, a left shift of 1 can be performed to remove the extended sign bit and store the result as a Q31 number.

To further understand a possible precision loss when manipulating Q-format numbers, let us consider another example where two Q3.12 numbers corresponding to 7.5 and 7.25 are multiplied and that the available memory space is 16-bit wide. As can be seen from Figure 5.5, the resulting product might be left shifted by 4 bits to store all the fractional bits corresponding to Q3.12 format. However, doing so results in a product value of 6.375, which is different than

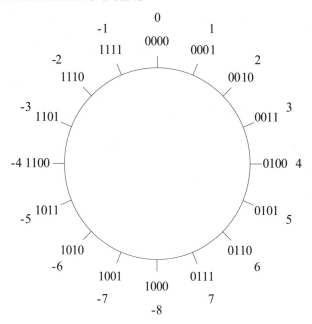

Figure 5.3: Four-bit binary representation.

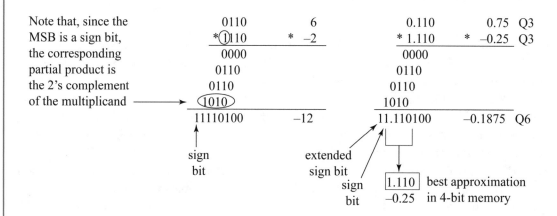

Note that, since the MSB is a sign bit, the corresponding partial product is the 2's complement of the multiplicand ⟶

Figure 5.4: Binary and fractional multiplication.

the correct value of 54.375. If the fractional product is stored in a lower precision Q-format—say, in Q6.9 format—then the correct product value can be stored.

Although Q-format solves the problem of overflow in multiplication, addition, and subtraction still pose a problem. When adding two Q15 numbers, the sum exceeds the range of

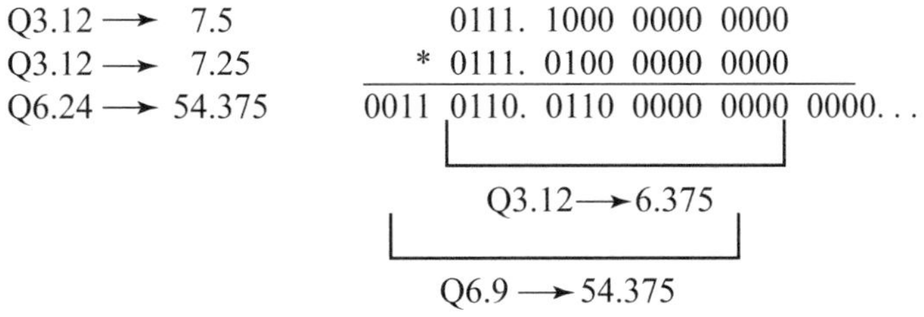

Q3.12 → 7.5   0111. 1000 0000 0000
Q3.12 → 7.25   * 0111. 0100 0000 0000
Q6.24 → 54.375   0011 0110. 0110 0000 0000 0000...

Q3.12 → 6.375

Q6.9 → 54.375

Figure 5.5: Q-format precision loss example.

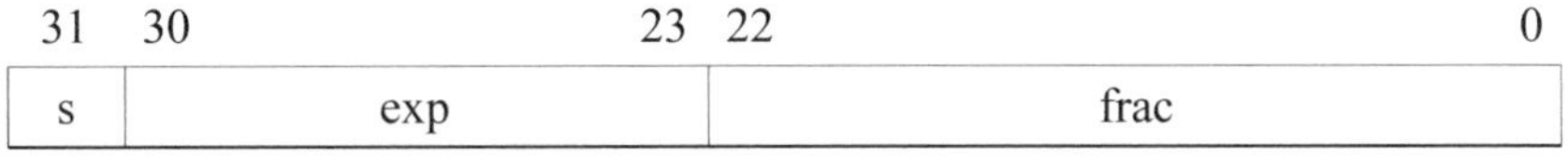

31	30		23	22		0
s	exp			frac		

Figure 5.6: Floating point data representation.

Q15 representation. To solve this problem, the scaling approach, discussed later in the chapter, needs to be employed.

5.2   FLOATING-POINT NUMBER REPRESENTATION

Due to relatively limited dynamic ranges of fixed-point processors, when using such processors, one should be concerned with the scaling issue, or how big the numbers get in the manipulation of a signal. Scaling is not an issue when using floating-point processors, since the floating-point hardware provides a much wider dynamic range. There are two floating-point data representations commonly in use: single-precision (SP) and double-precision (DP). In the SP format, a value is expressed as

$$-1^s * 2^{(exp-127)} * 1 \cdot frac, \tag{5.3}$$

where s denotes the sign bit (bit 31), exp the exponent bits (bits 23–30), and $frac$ the fractional or mantissa bits (bits 0–22); see Figure 5.6.

Consequently, numbers as big as $3.4 * 10^{38}$ and as small as $1.175 * 10^{-38}$ can be processed. In the double-precision format, more fractional and exponent bits are used as indicated:

$$-1^s * 2^{(exp-1023)} * 1 \cdot frac, \tag{5.4}$$

where the exponent bits are from bits 20–30 and the fractional bits are all the bits of one word and bits 0–19 of the other word; see Figure 5.7. In this manner, numbers as big as $1.7 * 10^{308}$ and as small as $2.2 * 10^{-308}$ can be handled.

When using a floating-point processor, all the steps needed to perform floating-point arithmetic are done by a floating-point CPU hardware. For example, consider adding two

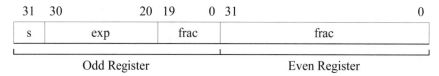

Figure 5.7: Double-precision floating-point representation.

floating-point numbers represented by

$$a = a_{frac} * 2^{a_{exp}}$$
$$b = b_{frac} * 2^{b_{exp}}.$$

$$(5.5)$$

The floating-point sum c has the following exponent and fractional parts:

$$c = a + b$$
$$= \left(a_{frac} + \left(b_{frac} * 2^{-\left(a_{exp}-b_{exp}\right)}\right)\right) * 2^{a_{exp}} \quad \text{if} \quad a_{exp} \geq b_{exp}$$
$$= \left(\left(a_{frac} * 2^{-\left(b_{exp}-a_{exp}\right)}\right) + b_{frac}\right) * 2^{b_{exp}} \quad \text{if} \quad a_{exp} < b_{exp}.$$

$$(5.6)$$

These parts are computed by the floating-point hardware. This shows that, although possible, it is inefficient to perform floating-point arithmetic on fixed-point processors, since all the operations involved, such as those in the above equations, need to be performed by software.

5.3   OVERFLOW AND SCALING

As stated before, fixed-point processors have a much smaller dynamic range than their floating-point processors. The ARM core is a fixed-point engine, mostly 32-bit. When available or used on an ARM architecture, the ARM floating-point hardware VFP enables floating-point operations. The audio data received from a smartphone microphone are normally 16 bit wide. That is why the Q15 representation of numbers needs to be considered. An ARM processor multiplier can multiply two Q15 numbers and produce a 32-bit product. The product can then be stored in 32 bits and used in subsequent computations, or shifted back to 16 bits for outputting through a smartphone speaker.

When multiplying two Q15 numbers, which are in the range of −1 and 1, the resulting number will always be in the same range. However, when two Q15 numbers are added, the sum may fall outside this range, leading to an overflow. Overflows can cause major problems by generating erroneous results. When using a fixed-point processor, the range of numbers must be closely monitored and adjusted to compensate for overflows. The simplest correction method for overflows is scaling.

Scaling can be applied to most filtering and transform operations, where the input is scaled down for processing and the output is then scaled back up to the original size. An easy way to do scaling is by shifting. Since a right shift of 1 is equivalent to a division by 2, one can scale the input repeatedly by 0.5 until all overflows disappear. The output can then be rescaled back by the total scaling amount.

As far as FIR and IIR filters are concerned, it is possible to scale coefficients to avoid over-flows. Let us consider the output of a filter $y[n] = \sum_{k=0}^{N-1} h[k] * x[n-k]$, where h's denote the coefficients or the unit sample response terms and x's input samples. In case of IIR filters, for a large enough N, the terms of the unit sample response become so small that they can be ignored. Let us suppose that x's are in Q15 format (i.e., $|x[n-k]| \leq 1$). Therefore, $|y[n]| \leq \sum_{k=0}^{N-1} |h[k]|$. This means that, to insure no output overflow (i.e., $|y[n]| \leq 1$), the condition $\sum_{k=0}^{N-1} |h[k]| \leq 1$ must be satisfied. This condition can be satisfied by repeatedly scaling (dividing by 2) the coef-ficients or the unit sample response terms.

5.4   SOME USEFUL ARITHMETIC OPERATIONS

There are many operations, such as division, trigonometric functions, and square root, that need to be obtained through approximations as no dedicated hardware is available for computing them. Here a number of techniques for implementing these operations are stated. Clearly, it takes many instructions for such operations to get computed. It should also be noted that, in some applications, it is more efficient to implement such arithmetic operations by using lookup tables.

5.4.1   DIVISION

The floating-point NEON coprocessor in modern smartphones provides an instruction, named VRECP, which provides an estimate of the reciprocal of an input number [1]. The accuracy can be improved by using this instruction as the seed point $v[0]$ for the iterative Newton–Raphson algorithm expressed by this equation

$$v[n+1] = v[n] * (2.0 - x * v[n]), \tag{5.7}$$

where x is the value whose reciprocal is to be found. Accuracy is increased by each iteration of this equation. A portion of this equation, i.e., $(2.0 - x * v[n])$, can be computed in the NEON coprocessor via the instruction VRECPS [2]. A full iteration is then achieved by the multiplica-tion with the previous value. In other words, on the floating-point NEON coprocessor, division can be achieved by taking the reciprocal of the denominator and then by multiplying the recip-rocal with the numerator. Accuracy is increased by performing many iterations. More details of the above division approach are covered in [3].

5.4.2   SINE AND COSINE

Trigonometric functions such as sine and cosine can be approximated by using the Taylor series expansion. For sine, the following expansion can be used:

$$\sin(x) = x - \frac{x^3}{3!} + \frac{x^5}{5!} - \frac{x^7}{7!} + \frac{x^9}{9!} + \text{higher order.} \tag{5.8}$$

Clearly adding higher-order terms leads to more precision. For implementation purposes, this expansion can be rewritten as follows:

$$\sin(x) \cong x * \left(1 - \frac{x^2}{2*3}\left(1 - \frac{x^2}{4*5}\left(1 - \frac{x^2}{6*7}\left(1 - \frac{x^2}{8*9}\right)\right)\right)\right). \tag{5.9}$$

Similarly, for cosine, the following expansion can be used:

$$\cos(x) \cong 1 - \frac{x^2}{2} + \frac{x^4}{4!} - \frac{x^6}{6!} + \frac{x^8}{8!}$$
$$= 1 - \frac{x^2}{2}\left(1 - \frac{x^2}{3*4}\left(1 - \frac{x^2}{5*6}\left(1 - \frac{x^2}{7*8}\right)\right)\right). \tag{5.10}$$

Furthermore, to generate sine and cosine, the following recursive equations can be used:

$$\sin nx = 2\cos x * \sin(n-1)x - \sin(n-2)x$$
$$\cos nx = 2\cos x * \cos(n-1)x - \cos(n-2)x. \tag{5.11}$$

5.4.3   SQUARE ROOT

Square root $sqrt(y)$ can be approximated by the following Taylor series expansion considering that $y^{0.5} = (x+1)^{0.5}$:

$$sqrt(y) \cong 1 + \frac{x}{2} - \frac{x^2}{8} + \frac{x^3}{16} - \frac{5x^4}{128} + \frac{7x^5}{256}$$
$$= 1 + \frac{x}{2} - 0.5\left(\frac{x}{2}\right)^2 + 0.5\left(\frac{x}{2}\right)^3 - 0.625\left(\frac{x}{2}\right)^4 + 0.875\left(\frac{x}{2}\right)^5. \tag{5.12}$$

Here, it is assumed that x is in Q15 format. In this equation, the estimation error would be small for x values near unity. Hence, to improve accuracy in applications where the range of x is known, x can be scaled by a^2 to bring it close to 1 (i.e., $sqrt(a^2 x)$ where $a^2 x \cong 1$). The result should then be scaled back by $1/a$.

It is also possible to compute square root by using the Newton–Raphson algorithm. The NEON coprocessor provides an instruction for the Newton–Raphson iteration of the reciprocal of square root:

$$v[n+1] = v[n] * (3 - x * v[n] * v[n])/2. \tag{5.13}$$

This instruction is named VRSQRTE which is used to provide an estimate of $1/sqrt(x)$ [3]. VRSQRTS provides the recursive equation part $(3 - x * w[n])/2$, where $w[n] = v[n] * v[n]$. A multiplication operation with the previous $v[n]$ is then needed for one full iteration of the previous equation.

L5   LAB 5: FIXED-POINT AND FLOATING-POINT OPERATIONS

In this lab, a typical computation function is implemented using both floating-point and fixed-point arithmetic and the differences in the outcomes are compared. The function considered is division.

A division instruction exists on ARM processors, which is performed by the floating-point coprocessor VFP. This instruction is generally slower than a multiplication instruction in terms of the number of clock cycles it takes to complete the computation. For example, on an ARM Cortex-A9 processor, the timing for fixed-point division is 15 cycles and for floating-point division is 25 cycles. Division can be obtained by using the following inversion operation:

$$y/x = y * 1/x. \tag{L5.1}$$

The inversion $\frac{1}{x}$ can be approximated using the iterative Newton–Raphson algorithm. Consider $v[0] = 1$ to be an initial estimate. The Newton–Raphson iterative equation for the reciprocal operation can be written as

$$f(y) = x - 1/y = 0 \tag{L5.2}$$
$$v(i + 1) = v(i) - (f(v(i)))/(f'(v(i))) \tag{L5.3}$$
$$= v(i) - (x - 1/v(i))/\left(-1/[v(i)]^2\right) \tag{L5.4}$$
$$= v(i) + (1 - x * v(i)) * v(i) \tag{L5.5}$$
$$= v(i) * (2x * v(i)). \tag{L5.6}$$

To double the accuracy, inputs can be scaled $(1 < x < 2)$ by appropriately moving the (binary) decimal point in x and y. This scaling needs to be removed to produce the final output.

L5.1   APP STRUCTURE

The base code provided in the previous labs is changed here as the real-time data processing is not the focus; rather, the accuracy of the output is the focus. A shell is provided in which C codes can get inserted. An option named Iteration is included in the *Settings* menu of the app to control the number of Newton–Raphson iterations that are to be performed to compute the reciprocal. The iteration needs to be implemented as a loop accepting the iteration variable as the number of times the loop will run.

L5.2   NEON SIMD COPROCESSOR

The NEON coprocessor capability can be accessed by either using assembly instructions or by using C intrinsic functions. Here, intrinsics are used. NEON support is available when targeting more recent ARM processors. When using an Android target, this is done by setting `armeabi-v7a` in the *abiFilter* directive and `-mfpu=neon` in the *cFlags* directive of the ndk section of the main *build.gradle* file. NEON intrinsics can then be used by adding the header `arm_neon.h` to the list of imports in your code. On iOS targets, the only step required to use NEON is adding the `arm_neon.h` header to the list of imports. NEON is a vector-based coprocessor, on which vectors can be processed. For each element, or lane, of a vector, the same operation is performed on all the elements. A listing of NEON intrinsics is available in [4].

The following example shows the procedure which performs a multiply-subtract operation using the NEON coprocessor intrinsics:

```
float32x4_t operandA;                    //quadword register
float32x4_t operandB;                    //quadword register
float32x4_t operandC;                    //quadword register
float32x4_t temp;                        //quadword register
float32_t neonResult[4] = {0,0,0,0};     //result vector
float32_t neonInputA[4] = {1.0, 2.0, 3.0, 4.0};      //input vector
float32_t neonInputB[4] = {5.0, 6.0, 7.0, 8.0};      //input vector
float32_t neonInputC[4] = {9.0, 10.0, 11.0, 12.0}; //input vector
operandA = vld1q_f32(neonInputA);        //load A into neon quadword register
operandB = vld1q_f32(neonInputB);        //load B into neon quadword register
operandC = vld1q_f32(neonInputB);        //load B into neon quadword register

temp = vmlsq_f32(operandA, operandB, operandC);   //compute temp=A-B*C
vst1q_f32(neonResult, temp);             //write back the result
```

In the above code, the variables of type `float32x4_t` refer to NEON registers. The type specifies that the register holds four 32-bit floating-point numbers. Since there is a total of 128 bits in the registers, they are referred to as quadword (Q) registers. NEON registers containing 64 bits are referred to as doubleword (D) registers. The NEON register bank is described in more detail in [5] and later in Chapter 9.

If the instructions are for operating on quadword registers, the suffix "q" is required to be added to the instruction intrinsic (as indicated above), otherwise, the registers will be assumed to be doubleword. The data type of the instruction needs to be specified as an additional _{type} suffix to the instruction. Supported types include 8-, 16-, 32-, and 64-bit signed and unsigned integers, as well as 32-bit floating-point. A complete listing of data types is available in [6].

L5.3   LAB EXERCISES

1. Write a fixed-point C code to implement the Newton–Raphson iterative equation. Consider Q15 format for the numbers. Use 16-bit variables for storage. Make sure appropriate amount of scaling/shifting is done to generate the highest precision Q15 format result. Compare the outcome with the outcome using a calculator. Next, adjust the number of Newton–Raphson iterations and explain the effect on the approximated result. What is the best tradeoff between number of iterations and accuracy of the resulting approximation? Examine the effect of lowering the Q format on the outcome. Keep in mind that Q2.13 format needs to be used for the number "2". Change your code to use 32-bit variables, choose an appropriate Q format, and repeat the above steps.

2. The NEON coprocessor is useful for optimizing performance intensive vector computations. The division operation mentioned earlier is a VFP instruction supporting only one set of operands. Reciprocal estimation and Newton–Raphson iteration functions are available on the NEON coprocessor [7, 8]. A reference implementation of the Newton–Raphson division using the NEON coprocessor is provided in the book software package. Compare the initial estimation from *VRECPE* with the double-precision division result (the reference case) from MATLAB, then add the *VRECPS* iteration and compare the results as the number of iterations is increased.

3. Write a floating-point C code to compute division by square root for two numbers b and a by making use of NEON intrinsics. You should use the provided reference code for Newton–Raphson division as the basis for your implementation. Compare the initial estimation with the result produced from the square-root operation performed in MATLAB. Add iterations as before and evaluate the outcome as the number of iterations is increased. How does the rate of convergence for square-root approximation compare to the rate of convergence for the division approximation?

5.6   REFERENCES

[1] http://infocenter.arm.com/help/index.jsp?topic=/com.arm.doc.dui0204j/CIHCHECJ.html 91

[2] http://infocenter.arm.com/help/index.jsp?topic=/com.arm.doc.dui0204j/CIHDIACI.html 91

[3] http://infocenter.arm.com/help/index.jsp?topic=/com.arm.doc.faqs/ka14282.html   91, 93

[4] http://infocenter.arm.com/help/topic/com.arm.doc.dui0491c/BABIIBBG.html 94

[5] http://infocenter.arm.com/help/index.jsp?topic=/com.arm.doc.dht0002a/ch01s03s02.html 94

[6] http://infocenter.arm.com/help/topic/com.arm.doc.dui0473c/CIHDIBDG.html 94

[7] http://infocenter.arm.com/help/topic/com.arm.doc.dui0204j/CIHCHECJ.html 95

[8] http://infocenter.arm.com/help/topic/com.arm.doc.dui0204j/CIHDIACI.html 95

CHAPTER 6

Real-Time Filtering

For carrying out real-time filtering, it is required to know how to acquire input samples, process them, and provide the result. This chapter addresses these issues toward achieving real-time filtering implementation on the ARM processor of a smartphone.

6.1 FIR FILTER IMPLEMENTATION

An FIR filter can be implemented in C using standard math functions or by using NEON intrinsics. The goal of the implementation is to have a minimum cycle time algorithm. This means that performing the filtering as fast as possible in order to achieve the highest sampling frequency (the smallest sampling time interval). Initially, the filter is implemented in C, since this demands the least coding effort. In Chapter 9 covering optimization, alternative implementations will be considered based on this algorithm. The difference equation $y[n] = \sum_{k=0}^{N-1} B_k * x[n-k]$ is implemented to realize the filter.

Considering Q15 representation here, 16-bit samples need to get multiplied by 16-bit coefficients. In order to store the product in 32 bits, it has to be left shifted by one to get rid of the extended sign bit. Now, to export the product to the output, it must be right shifted by 16 to place it in the lower 16 bits. Alternatively, the product may be right shifted by 15 without removing the sign bit but this will result in a slight loss of precision in the intermediate result accumulator.

To implement the algorithm in C, standard multiplication expressions and the shift operators "<<" and ">>" are used as shown below. Here, `result` is 32 bits wide, while the variables `coefficient` and `sample` are 16 bits wide.

```
result = ( coefficient * sample ) << 1;
result = result >> 16;
```

or

```
result = ( coefficient * sample ) >> 15;
```

For the proper operation of the FIR filter, it is required that the current sample and $N - 1$ previous samples be processed at the same time, where N is the number of coefficients. Hence, the N most current samples have to be stored and updated with each incoming sample. This can be done easily via the following code.

```
static void insertNewSample(short inSample) {
    int i;
    // Update array samples
    for(i=0;i<N-1;i++){
        samples[i] = samples[i+1];
    }
    samples[N-1] = inSample;
}
```

Here, as a new sample comes in, each of the samples is moved into the previous location in the array. As a result, the oldest sample `sample[0]` is discarded, and the newest sample is put into the array location `samples[N-1]`.

Now that the N most current samples are in the array, the filtering operation may get started. All that needs to be done, according to the difference equation, is to multiply each sample by the corresponding coefficient and sum the products. This is achieved by the following code.

```
static short processFIRFilter(short inSample) {
    int i, result = 0;

    // Update array samples
    for( i = 0; i < N; i++ ){
        samples[i] = samples[i+1];
    }
    samples[N-1] = inSample;

    // Filtering
    for( i = 0; i < N; i++ ){
        result += (samples[i] * coefficients[i]) ) << 1;
    }
    return ( result >> 16);
}
```

This approach adds some overhead to the computation because of the memory changes. Memory accesses are very expensive in terms of processing time, and having many such accesses is not computationally efficient for implementation on ARM processors. It should be noted that the proper way of doing this type of filtering is by using circular buffering. This approach is discussed in the next section.

6.2   CIRCULAR BUFFERING

In many signal processing algorithms, including filtering, adaptive filtering, and spectral analysis, one requires to shift data or update samples, or to deal with a moving window. For example, an FIR filter basically consists of a list of coefficients which are multiplied with current and past sampled input signal values. FIR filtering does not involve dependency on previous output values. For a filter with N coefficients, the output $y(n)$ based on the input signal $x(n)$ gets computed via an FIR filter, that is $y(n) = \sum_{k=0}^{N-1} B_k * x(n-k)$. This equation can be realized by using a circular buffer where a moving window of the past $N-1$ sampled values and the current sampled value is established to compute the output corresponding to the current input sample.

The direct method of shifting data is normally not efficient and uses many cycles. Circular buffering is an addressing mode by which a moving-window effect can be created without the overhead associated with data shifting. In a circular buffer, if a pointer pointing to the last element of the buffer is incremented, it is automatically wrapped around and pointed back to the first element of the buffer. This provides an easy mechanism to exclude the oldest sample while including the newest sample, creating a moving-window effect as illustrated in Figure 6.1.

Some processors such as a DSP processor have specialized hardware for circular buffering. However, on ARM processors, such hardware is not available. Here, let us consider a C implementation of a circular buffer which could be used on an ARM processor. First, let us establish the data structure of a circular buffer:

```
typedef struct CircularBuffer {
          short* buff;
          int writeCount;
          int bitMask;
} CircularBuffer;
```

In the `CircularBuffer` structure, `buff` contains the data and `writeCount` stores the number of times the circular buffer has been written to. The rest of the code for the circular buffer appears below.

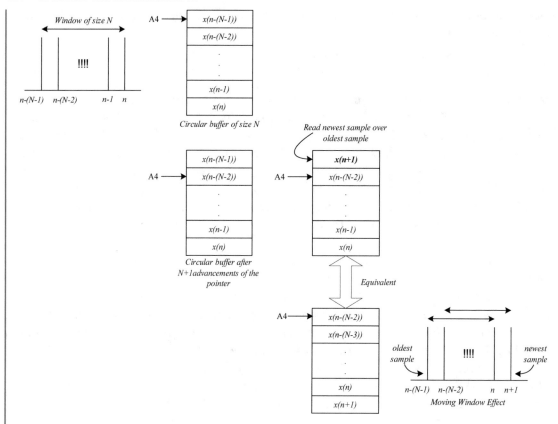

Figure 6.1: Moving window effect.

```
CircularBuffer* newCircularBuffer(int size) {
    CircularBuffer* newCircularBuffer =
                (CircularBuffer*)malloc(sizeof(CircularBuffer));

    int pow2Size = 0x01;
    while (pow2Size < size) {
        pow2Size = pow2Size << 1;
    }

    newCircularBuffer->writeCount = 0;
    newCircularBuffer->bitMask = pow2Size-1;
    newCircularBuffer->buff = (short*)calloc(pow2Size,sizeof(short));
    return newCircularBuffer;
```

```
}

void writeCircularBuffer(CircularBuffer* buffer, short value){
    //buff[writeCount % pow2Size] = value;
    //writeCount = writeCount + 1;
    buffer->buff[(buffer->writeCount++) & buffer->bitMask] = value;
}

short readCircularBuffer(CircularBuffer* buffer, int index){
    //return buff[(writeCount - ( index + 1 )) % pow2Size]
    return buffer->buff[(buffer->writeCount + (~index))
            & buffer->bitMask];
}

void destroyCircularBuffer(CircularBuffer** buffer){
    free((*buffer)->buff);
    (*buffer)->buff = NULL;
    free(*buffer);
    *buffer = NULL;
}
```

Since there is not a specialized hardware for computing the current address in circular buffers on ARM processors, alternative methods typically involve the modulus operation. The use of the `bitMask` field is important since a special case of the modulus operation is used here where the `size` of the circular buffer is a power of two. The modulus operation inherently involves the division operation which is typically a slow instruction to execute. Instead, a bit mask is stored. When the index is bitwise ANDed with the bit mask, only the lower bits of the index remain—effectively the same result that a modulo operation would generate.

The methods to interface with the `CircularBuffer` structure consist of `newCircularBuffer` which initializes the memory locations and the corresponding `destroyCircularBuffer` which releases the memory locations. The `writeCircularBuffer` method inserts the sample given by `value` into `buff` and increments the `writeCount`. Last, `readCircularBuffer` returns the value at `index`, where `index` zero corresponds to the newest sample inserted into `buff` and increasing values corresponding to samples further in the past. An implementation of this circular buffer is shown in the following code.

```
static short processFIRFilter(short inSample){
    writeCircularBuffer(inputBuffer, inSample);
```

```
      int i, result = 0;
      for( i = 0; i < N; i++) {
          result += (readCircularBuffer(inputBuffer, i)
                                  * coefficients[i]) << 1;
      }
      return (result >> 16);
}
```

Although the performance of this method may appear acceptable, it is not the best that can be achieved on ARM processors. To reduce the computational burden of creating a moving window of input samples, an alternative approach known as frame processing is discussed next.

6.3   FRAME PROCESSING

When it comes to processing frames of data, such as when performing FFT or block convolution, triple buffering is an efficient data frame handling mechanism. While samples of the current frame are being collected by the CPU in an input buffer, samples of the previous frame in an intermediate array can get processed during the time left between frame captures. At the same time, the samples of a previously processed frame available in an output array can be sent out. In this manner, the CPU is used to set up the input array and process the intermediate array while processed data is moved from the output array. At the end of each frame or the start of a new frame, the roles of these arrays can be interchanged. The input array is reassigned as the intermediate array to be processed, the processed intermediate array is reassigned as the output array to be sent out, and the output array is reassigned as the input array to collect incoming samples for the current frame. In the FIR filtering method mentioned above, it is required that one keeps a memory of previous signal values. If processing is performed on framed segments of N input signal samples, one requires access to $N + 1 + M$ sample values. This process is illustrated in Figure 6.2.

Depending on the size of the filter M, the overlap could require more than three frames to be kept. Using the technique of frame processing, it is possible to reduce the computational burden of creating a windowed signal by allowing chunks of input data to be sampled, processed, and stored at once instead of performing all of these operations for each sample individually. The following code shows how an input frame of data can be processed using the frame processing method.

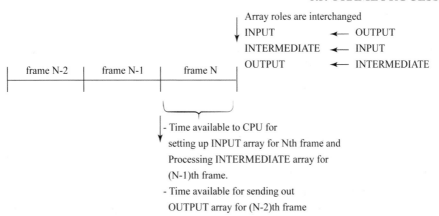

Figure 6.2: Triple buffering technique.

```
static void processFIRFilter(short* inputSamples, short* output) {
    int i, j;
    for( i = 0; i < N; i++) {
        inputBuffer[i] = inputBuffer[N + i];
        inputBuffer[N + i] = inputSamples[i];
    }
    for( i = 0; i < N; i++) {
        temp = 0;
        for( j = 0; j < N; j++) {
            idx = N + (i - j);
            result += (inputBuffer[idx]*coefficients[j])<<1;
        }
        output[i] = (short) (result>>16);
    }
}
```

In the above code, the input frame size is considered to be equal to the number of coefficients in the FIR filter. Thus, `inputBuffer`, which is an array of type *short*, should be twice the number of FIR coefficients. Index zero in the array corresponds to the oldest sample and index $2N - 1$ corresponds to the newest sample. The first "for loop" performs the operation of shifting the previous frame to the lower half of the array while copying in new samples to the upper half. The second "for loop" implements the FIR convolution. Since `inputBuffer` contains the newest samples in its upper half, the `idx` expression is computed to get the correct sample corresponding to the FIR filter coefficient.

6.4   FINITE WORD LENGTH EFFECT

Due to the fact that memory or registers have finite number of bits, there could be a noticeable error between desired and actual outcomes on a fixed-point processor. The so-called finite word length quantization effect is similar to input data quantization effect introduced by an A/D converter.

Consider fractional numbers quantized by a $b + 1$ bit converter. When these numbers are manipulated and stored in an $M + 1$ bit memory, with $M < b$, there is going to be an error (simply because $b - M$ of the least significant fractional bits are discarded or truncated). This finite word length error could alter the behavior of a system to an unacceptable degree. The range of the magnitude of truncation error ϵ_t is given by $0 \leq |\epsilon_t| \leq 2^M - 2^b$. The lowest level of truncation error corresponds to the situation when all the thrown-away bits are zeros, and the highest level to the situation when all the thrown-away bits are ones.

This effect has been extensively studied for FIR and IIR filters (for example, see [1]). Since the coefficients of such filters are represented by a finite number of bits, the roots of their transfer function polynomials, or the positions of their zeros and poles, shift in the complex plane. The amount of shift in the positions of poles and zeros can be linked to the amount of quantization error in the coefficients.

Also, note that as a result of coefficient quantization, the actual frequency response $\hat{H}\left(e^{j\theta}\right)$ would become different than the desired frequency response $H\left(e^{j\theta}\right)$. For example, for an FIR filter having N coefficients, it can be easily shown that the amount of error in the magnitude of the frequency response, $\left|\Delta H\left(e^{j\theta}\right)\right|$, is bounded by

$$\left|\Delta H\left(e^{j\theta}\right)\right| = \left|H\left(e^{j\theta}\right) - \hat{H}\left(e^{j\theta}\right)\right| \leq N 2^{-b}. \tag{6.1}$$

L6   LAB 6:
REAL-TIME FIR FILTERING, QUANTIZATION EFFECT, AND OVERFLOW

The purpose of this lab is to design and then run an FIR filter written in C on the ARM processor of a smartphone. Also, the quantization effect and overflow are examined. The application shells introduced in the previous labs are used here for collecting audio signal, passing a sampled signal to a C code segment for filtering, and saving the output to a file for analysis. The design of the FIR filter, i.e., generation of filter coefficients, is realized using the MATLAB filter design tool. Other filter design tools may be used for the generation of filter coefficients. The lab experiment involves the implementation of this filter in C code. The base application shell is used here to insert the filtering C code. Note that the previous lab provided a floating-point C code implementation of FIR filtering while the fixed-point C code implementation of FIR filtering is covered here.

L6.1   FILTER DESIGN

To generate the filter coefficients, the Parks-McClellan method is used to design a lowpass filter for the specifications stated below.

```
rpass = 0.1;                    %passband ripple
rstop = 20;                     %stopbad ripple
fs = 44100;                     %sampling frequency
f = [1600 2400];                %cutoff frequencies
a = [1 0];                      %desired amplitudes


%compute deviations
dev = [(10^(rpass/20)-1)/(10^(rpass/20)+1) 10^(-rstop/20)];


[n,fo,ao,w] = firpmord(f,a,dev,fs);  %estimate filter parameters
B = firpm(n,fo,ao,w);                %generate filter coefficients
```

This code creates an array of $n + 1$ double-precision filter coefficients meeting the above specifications. In order to confirm that the filter matches the specifications, a synthesized signal is considered for testing via the following MATLAB code (see Figure 6.3). This code synthesizes a sinusoidal signal composed of three frequency components. The signal gets filtered and the spectrum of the input and output signals are displayed along with the frequency response of the filter.

```
ts = 1/fs;        %sample time
ns = 512;         %number of fft points
t = [0:ts:ts*(ns-1)];


%generate test signal to verify filter
f1 = 750;
f2 = 2500;
f3 = 3000;
x = sin(2*pi*f1*t) + sin(2*pi*f2*t) + sin(2*pi*f3*t);


%plot fft of synthesized signal
X = abs(fft(x,ns));
X = X(1:length(X)/2);
frq = [1:1:length(X)];
subplot(3,1,1);
plot(frq*(fs/ns),X);
```

```
grid on;

%plot normalized frequency of filter
A = 1;
[h,w] = freqz(B,A,512);
subplot(3,1,2);
plot(w/(2*pi),10*log(abs(h)));
grid on;

%plot fft of filtered signal
Y = filter(B,A,x);
Y = abs(fft(Y,ns));
Y = Y(1:length(Y)/2);
frq = [1:1:length(Y)];
subplot(3,1,3);
plot(frq*(fs/ns),Y);
grid on;
```

L6.2   ARM OVERFLOW DETECTION

A method for detecting when an overflow occurs can be implemented by using ARM arithmetic instructions which update the Application Program Status Register (APSR). By performing an operation using the arithmetic instruction ADDS, the APSR can then be read to determine overflows. The following code shows an implementation of the ADDS instruction in ARM assembly which uses this register.

```
#ifdef __arm__
        .align      2
        .global     addStatus
addStatus:
            @ r0 = input A and sum output address
            @ r1 = input B and status output address
        ldr    r2,   [r0]          @ load contents at [r0] into r2
        ldr    r3,   [r1]          @ load contents at [r1] into r3
        adds   r2,   r2,   r3      @ add r2 and r3, store in r2
                                   @ set status flags in APSR
        mrs    r3,   APSR          @ copy APSR to r3
        str    r2,   [r0]          @ store r2 at address [r0]
```

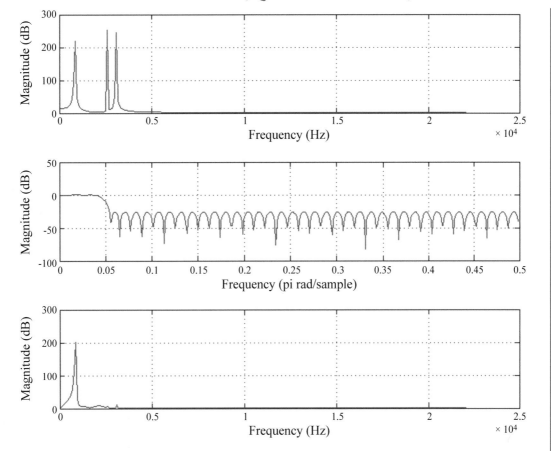

Figure 6.3: Synthesized signal.

```
        str    r3,    [r1]              @ store r3 at address [r1]
        bx     lr                       @ return to caller
#elif __arm64__
        .align    2
        .global   addStatus
addStatus:
        @ x0 = input A and sum output address
        @ x1 = input B and status output address
        ldr    w2,    [x0]              @ load contents at [x0] into w2
        ldr    w3,    [x1]              @ load contents at [x1] into w3
        adds   w2,    w2,    w3         @ add w2 and w3, store in w2
```

```
                                    @ set status flags in APSR
      mrs    x3,    NZCV            @ copy APSR to x3
      str    w2,    [x0]            @ store w2 at address [x0]
      str    w3,    [x1]            @ store x3's MSB at address [r1]
      ret
#endif
```

Since ARMv7 and ARMv8 instruction sets are used on different iOS devices, the assembly code to check the APSR needs to be implemented properly for each instruction set. The line `#ifdef __arm__` checks at compile time if the supported instruction set is ARMv7. If the instruction set is not ARMv7, it checks if the instructions set is ARMv8 using the __arm64__ flag. There are some notable differences between the two implementations. ARMv8 uses 64-bit registers, thus the register naming in the assembly code is different. On ARMv8, the register file consists of 64-bit segments R0–R30. When used in assembly coding, these registers must be further qualified to indicate the operand data size. Registers beginning with X refer to full-width 64-bit registers whereas registers beginning with W refer to 32-bit registers. Also, note that the mnemonic for the status register is APSR on ARMv7 and NZCV on ARMv8. The assembly code segment can then be called from C in the manner shown below.

```
static inline int
addGetStatus(short a, short b, short *c) {
      //performs the operation a+b=c and returns the status register
      int A = a << 16;
      int B = b << 16;
      addStatus(&A,&B);
      (*c) = A >> 16;
      return B;
}
```

After the execution of addGetStatus, the register A contains the status register and B contains the result of the addition. The following test cases illustrate the operation of the addStatus function.

On an Android platform, `__android_log_print` is used to send the output to LogCat:

```
unsigned int status;
short result;

short A = 32767;
```

```
short B = 32767;
status = addGetStatus(A, B, &result);
__android_log_print(ANDROID_LOG_ERROR, ''Add Status'',
        ''A: %d, B: %d, C: %d, Status: %#010x'', A, B, result, status);


A = 32767;
B = -32768;
status = addGetStatus(A, B, &result);
__android_log_print(ANDROID_LOG_ERROR, ''Add Status'',
        ''A: %d, B: %d, C: %d, Status: %#010x'', A, B, result, status);


A = 10;
B = 11;
status = addGetStatus(A, B, &result);
__android_log_print(ANDROID_LOG_ERROR, ''Add Status'',
        ''A: %d, B: %d, C: %d, Status: %#010x'', A, B, result, status);


A = -10;
B = -10;
status = addGetStatus(A, B, &result);
__android_log_print(ANDROID_LOG_ERROR, ''Add Status'',
        ''A: %d, B: %d, C: %d, Status: %#010x'', A, B, result, status);


A = 100;
B = -1000;
status = addGetStatus(A, B, &result);
__android_log_print(ANDROID_LOG_ERROR, ''Add Status'',
        ''A: %d, B: %d, C: %d, Status: %#010x'', A, B, result, status);


A = -100;
B = -32768;
status = addGetStatus(A, B, &result);
__android_log_print(ANDROID_LOG_ERROR, ''Add Status'',
        ''A: %d, B: %d, C: %d, Status: %#010x'', A, B, result, status);


A = 32767;
B = 1000;
status = addGetStatus(A, B, &result);
```

```
__android_log_print(ANDROID_LOG_ERROR, ''Add Status'',
      ''A: %d, B: %d, C: %d, Status: %#010x'', A, B, result, status);
```

On an iOS platform, the same output is shown using the `printf` method:

```
unsigned int status;
short result;

short A = 32767;
short B = 32767;
status = addGetStatus(A, B, &result);
printf(''A: %d,B: %d,C: %d,Status: %#010x\n'', A, B, result, status);

A = 32767;
B = -32768;
status = addGetStatus(A, B, &result);
printf(''A: %d,B: %d,C: %d,Status: %#010x\n'', A, B, result, status);

A = 10;
B = 11;
status = addGetStatus(A, B, &result);
printf(''A: %d,B: %d,C: %d,Status: %#010x\n'', A, B, result, status);

A = -10;
B = -10;
status = addGetStatus(A, B, &result);
printf(''A: %d,B: %d,C: %d,Status: %#010x\n'', A, B, result, status);

A = 100;
B = -1000;
status = addGetStatus(A, B, &result);
printf(''A: %d,B: %d,C: %d,Status: %#010x\n'', A, B, result, status);

A = -100;
B = -32768;
status = addGetStatus(A, B, &result);
printf(''A: %d,B: %d,C: %d,Status: %#010x\n'', A, B, result, status);
```

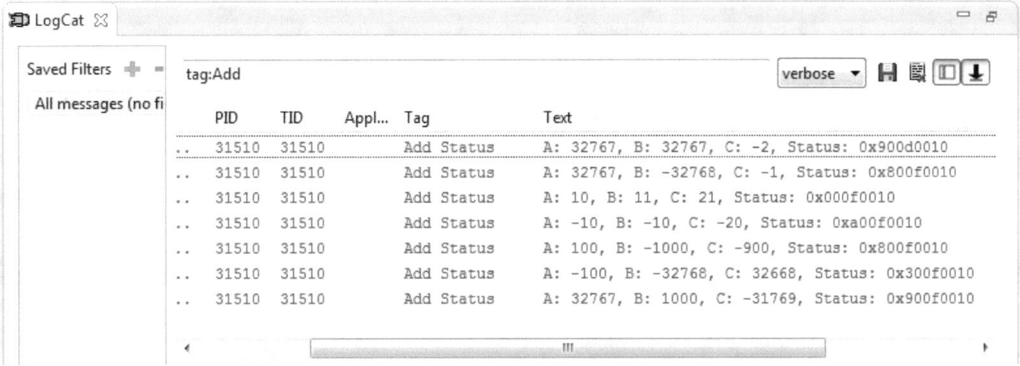

Figure 6.4: LogCat screen.

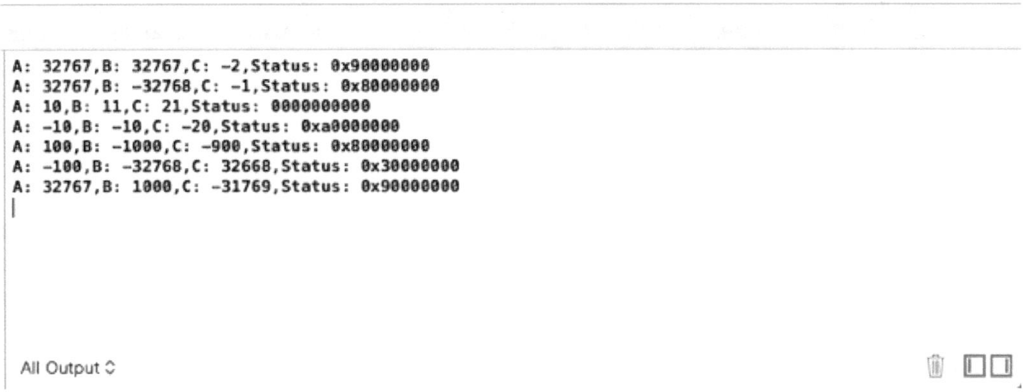

Figure 6.5: Output values.

```
A = 32767;
B = 1000;
status = addGetStatus(A, B, &result);
printf(''A: %d,B: %d,C: %d,Status: %#010x\n'', A, B, result, status);
```

The first half-byte of the `status` word copied from the APSR contains the NZCV (negative, zero, carry, and overflow) bit flags. The outcome from the test cases is shown in Figures 6.4 and 6.5. The first hexadecimal character corresponds to the bits of the NZCV flags. For the case of $2^{15} - 1 + 2^{15} - 1$ (the largest positive value represented by Q15 numbers), one can see the resulting status of 0x9 or binary 1001. This means that the result became negative and produced an overflow.

L6.3   LAB EXERCISES

1. Use the MATLAB code below to design an FIR filter based on the specifications noted. This filter is obtained using a different design method than the one mentioned above via the *fir2()* function of MATLAB. Refer to the MATLAB documentation on how to use this function.

```
f = [0 600 1000 1400 2000 4000];    %frequencies (0 to nyquist)
a = [1.25 2 1 .25 0.1 0];           %desired amplitude response
order = 31;                         %filter order N
```

Next, quantize the filter by using the MATLAB fixed-point toolbox function *sfi()*. For example, if `coeffs` denotes double-precision filter coefficients, `ficoeffs = sfi(coeffs,bits,bits-intgr-1)` can be used to convert to quantized values with `bits` denoting wordlength, `intgr` integer bits, and `intgr-1` fractional bits. If the magnitude of any of the coefficients is greater than or equal to 1, an appropriate amount of integer bits needs to be used. To retrieve quantized coefficients, the function `ficoeffs.data` can be used.

Test your filter using the signal: `chirp(t,0,ts*ns,fs/2)` with `ns` equal to 256 samples. Determine the minimum number of bits needed to represent the coefficients such that the maximum absolute error of the filter in the frequency domain is less than 5%, i.e., the comparison should be between the frequency spectrum of the filtered output based on the quantized filter output and the double precision unquantized filter output.

2. Implement the filter designed above in C using 16-bit short values for the coefficients as well as for the input samples. All computations are to be performed using fixed-point representation.

The filter output may get overflowed due to the multiplications and summations involved in the FIR filtering equation. Develop a scheme to detect such overflows and implement a prevention measure. The most effective way to avoid overflows is by scaling down the input signal magnitude before filtering and then reversing the scaling when the output is returned. Keep scaling the input signal samples by a scaling factor less than one (a scaling factor of 1/2 can be achieved simply by right shifting) until the overflow disappears.

6.6   REFERENCES

[1] J. Proakis and D. Manolakis, *Digital Signal Processing: Principles, Algorithms, and Applications*, Prentice Hall, 1996. 104

CHAPTER 7

Adaptive Filtering

In this chapter, an adaptive FIR filter is used to model the behavior of an IIR filter. Let us first examine IIR filtering.

7.1 INFINITE IMPULSE RESPONSE FILTERS

An IIR filter has a unit sample response that is infinite in time because of its recursive dependence on previous output values. IIR filters are described according to the following difference equation:

$$y[n] = -\sum_{k=1}^{N} ak * y[n-k] + \sum_{k=0}^{N} bk * x[n-k], \qquad (7.1)$$

where a_k's and b_k's denote the coefficients. The recursive behavior of the filter is caused by the feedback provided from the a_k coefficients acting on the previous output terms $y[n-k]$. This is called Direct Form I and the following C code implements it:

```
static float infiniteIR(float inSample) {
    int i;
    float aSum, bSum;

    for( i = 0; i < N; i++ ){
        x[i] = x[i+1];
    }
    x[N-1] = inSample;

    bSum = 0;
    for( i = 0; i < = N; i++) {
        bSum += x[N-i]*b[i];
    }

    aSum = 0;
    for( i = 1; i <= N; i++ ) {
        aSum += y[i-1]*a[i];
    }
```

```
for( i = N-1; i > 0; i--) {
    y[i] = y[i-1];
}
y[0] = bSum - aSum;

return y[0];
}
```

Compared to FIR filters, IIE filters have advantages and disadvantages. IIR filters allow meeting a desired frequency response characteristic via a lower number of coefficients than an equivalent FIR filter, resulting in a lower computation time. FIR filters provide linear phase response whereas IIR filters do not. Unlike FIR filters which have no poles and are thus guaranteed to be stable, special care must be taken with IIR filters. Stability of IIR filters is heavily dependent on quantization. Recall the finite word length effect discussed in previous chapters. The amount of shift in the positions of poles and zeros can be related to the amount of quantization error in the coefficients. For an Nth order IIR filter, the sensitivity of the ith pole p_i with respect to the kth coefficient a_k can be derived to be [1],

$$\frac{\partial p_i}{\partial A_k} = \frac{-p_i^{N-k}}{\prod_{\substack{l=1 \\ l \neq i}}^{N} (p_i - p_l)}.$$

(7.2)

This means that the change in the position of a pole is influenced by the positions of all the other poles. That is the reason an Nth order IIR filter is normally implemented by having a number of second-order IIR filters in series in order to decouple this dependency of poles and having real value coefficients.

7.2   ADAPTIVE FILTERING

Adaptive filtering is used in many applications such as noise cancellation and system identification. The most common method of adaptive filtering is to modify the coefficients of an FIR filter according to an error signal to adapt to a desired signal. In system identification, the behavior of an unknown system is modeled by comparing the output $d[n]$ of the unknown system with the output $y[n]$ of a known system, in our case an adaptive FIR filter. The difference in the output of the unknown system and the adaptive FIR filter is considered to be the error term $e[n]$, which is used to update the coefficients of the adaptive filter. This process is illustrated in Figure 7.1.

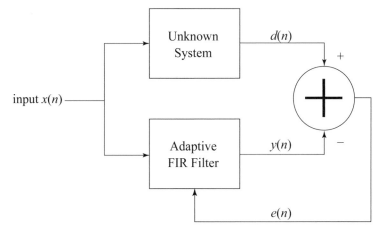

Figure 7.1: System identification by adaptive filtering.

The LMS algorithm is widely used to update each coefficient of the FIR filter according to the following equation:

$$h_n[k] = h_{n-1}[k] + \delta * e[n] * x[n - k], \qquad (7.3)$$

where h's denote the FIR filter coefficients. The output $y[n]$ converges to $d[n]$. The rate of convergence is governed by the step size δ. Small step sizes ensure convergence at the cost of slow adaptation rate. Larger step sizes lead to faster adaptation at the cost of overshooting the solution. Additionally, the order of the FIR filter used limits how accurately the unknown system can be modeled. A low-order filter will likely not be able to produce an accurate model. Conversely, a high-order filter might produce an accurate model but the computational complexity of such a filter would be prohibitive for real-time operation purposes.

L7   LAB 7:
IIR FILTERING AND ADAPTIVE FIR FILTERING

This lab consists of two parts. In the first part, the direct form realization of an IIR filter of order N is compared to its cascade form realization (i.e., $N/2$ second-order filters). In the second part, an adaptive FIR filter is used to model or match the IIR filter.

L7.1   IIR FILTER DESIGN

The unknown system that is used as the target for the adaptive FIR filter is an eight-order band pass IIR filter. The MATLAB function *yulewalk* is used here to achieve the desired filter design. Note that the frequency band definitions are decimal numbers ranging from 0–1, with 1 representing Nyquist frequency. The actual sampling rate of the input signal is not needed to generate the filter coefficients. The MATLAB code below is used to design the filter with the pass-band from $\pi/3$ to $2\pi/3$ radians and 20-dB stop-band attenuation:

```
Nc=8;                           %8th order IIR
f=[0 0.32 0.33 0.66 0.67 1];    %frequency band definition
m=[0 0 1 1 0 0];                %gain definition
[B,A]=yulewalk(Nc,f,m);         %coefficient computation
[h,w]=freqz(B,A);               %graph filter response
plot(f,m,w/pi,abs(h),'--');
legend('Ideal','yulewalk Designed')

%test signal parameters
Fs=8000;
Ts=1/Fs;
Ns=128;
t=[0:Ts:Ts*(Ns-1)];

%generate test signal to verify filter
f1 = 750;
f2 = 2000;
f3 = 3000;
x = 1/3*sin(2*pi*f1*t) + 1/3*sin(2*pi*f2*t) + 1/3*sin(2*pi*f3*t);

%filter the signal
y=filter(B,A,x);
```

The output of the filter can be verified by graphing the filter response in MATLAB as well as the frequency content of the sampled signal and the filtered signal (shown in Figures 7.2 and 7.3). The MATLAB code to produce these comparisons was previously shown in Lab L6.

When performing the implementation on a smartphone target, the stop- and pass-band frequencies and gains may need to be adjusted to hear an audible difference as the frequency response of the filter will scale along with the sampling rate used to record and process audio on the smartphone.

L7.2   ADAPTIVE FIR FILTER

The adaptive FIR filter implemented here is based on the LMS algorithm. It is designed to adapt to the output of an unknown system, which is considered to be an IIR filter. The following C code performs the adaptation to the output of the IIR filter.

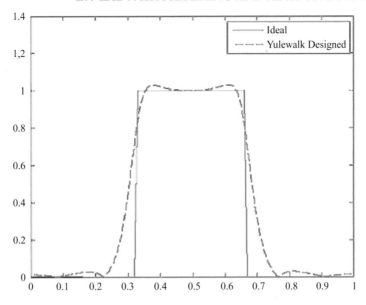

Figure 7.2: Bandpass filter.

```
float adaptiveFIR(float inSample) {
    int i;
    float error, weight;
    float mu = 0.01f;

    for( i = 0; i < N; i++ ){
        xf[i] = xf[i+1];
    }
    xf[N-1] = inSample;

    float firOutput = 0.0f;
    for( i = 0; i < FIRorder; i++) {
        firOutput += xf[FIRorder-i]* coefficients[i];
    }

    error = infiniteIR(inSample)-firOutput;
    weight = mu*error;

    for( i = 0; i < FIRorder; i++) {
```

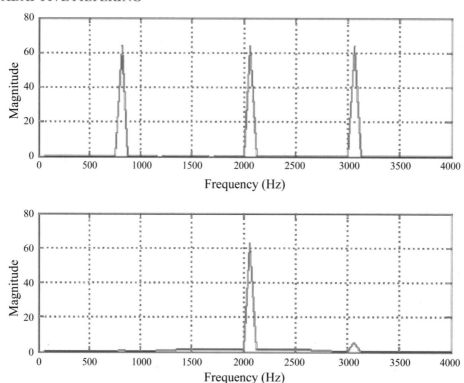

Figure 7.3: Filtered response.

```
            coefficients[i] += weight*xf[FIRorder-i];
        }
        return firOutput;
    }
```

When this method is initially called, the FIR filter uses a null set of filter coefficients. As each sample of a frame is filtered, the FIR coefficients get updated.

L7.3   LAB EXERCISES

1. Use MATLAB to design an 8-order order bandpass IIR filter based on the following specifications:

```
f=[0 0.2 0.4 0.41 1];        %frequency bands
m=[0 1 1 0 0];               %desired gain
```

Study the round-off error between the direct form and the second-order cascade form using MATLAB. Use the MATLAB function *tf2sos* to convert the transfer function into cascade form and then apply the filter *sos* as indicated below:

```
Z = x;
for i=1:size(sos);
    Z = filter(sos(i,1:3),sos(i,4:6), Z);
end
```

Examine the effect of various word lengths on the output and report your observations. Recall that you can quantize the filter by using the MATLAB fixed-point toolbox function `sfi()`. For example, if `coeffs` denotes double-precision filter coefficients, the expression `ficoeffs = sfi(coeffs,bits,bits-intgr-1)` can be used to convert to quantized values with `bits` denoting wordlength, `intgr` integer bits, and `intgr-1` fractional bits. Although this issue did not arise during the exercise involving FIR coefficient quantization, it needs to be noted that the number of integer bits must be sufficient to accommodate IIR filter coefficients whose magnitude is greater than 1.

First, compare the frequency spectra of the filtered outputs of the direct form filter based on the quantized and unquantized coefficients. Then, compare the frequency spectra of the filtered outputs based on the direct form quantized coefficients and the quantized second-order sections coefficients.

2. Over time, the output of the FIR filter should converge to that of the IIR filter. Confirm this by comparing the output of the two filters or by examining the decline in the error term. Experiment with different step size δ and filter length N and report your observations.

   Next, add a delay to the adaptive FIR filtering pipeline to make the real-time processing fail on purpose. A possible solution to address the real-time processing aspect is to update only a fraction of the coefficients using the LMS equation during each iteration. For example, you may update all even coefficients during the first iteration, and then all odd coefficients during the second iteration. Implement such a coefficient update scheme and report the results in terms of the tradeoff between convergence rate, convergence accuracy, and processing time.

Hints (Android target)   The function `clock_gettime` with `CLOCK_MONOTONIC` as the time source can be used to obtain high-resolution timing. An implementation of this function is provided in the Android lab source code in the form of a separate C object (Timer.c) that can be placed within your code to provide the necessary benchmark timing. Also, the AndroidPlot graphing library is used to display the filtering error term in real-time as an audio signal is

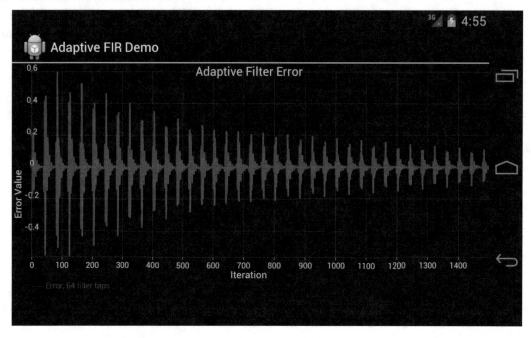

Figure 7.4: Adaptive FIR app.

getting processed. The AChartEngine library is also used to display the initial N-frame error term, as shown in Figure 7.4. Graphing implementations are found in AdaptiveFilter.java and ErrorGraphActivity.java and the dependencies are specified in the build.gradle file.

Hints (iOS target)   To measure the execution time taken by the signal processing pipeline, the header file mach_time.h can be used. This header measures time in the form of a tick count, which can be easily converted to nanoseconds. You can use this to find the execution time of your code in Objective-C and C. To measure the execution time taken by a C code, the time.h header file can be utilized. The function clock() defined in time.h can be used to calculate the execution time between any two places of the code.

7.4   REFERENCES

[1] J. Proakis and D. Manolakis, *Digital Signal Processing: Principles, Algorithms, and Applications*, Prentice Hall, 1996. 114

CHAPTER 8

Domain Transforms

Frequency domain transforms are extensively used in signal processing applications. In this chapter, the Discrete Fourier Transform (DFT) and the Fast Fourier Transform (FFT), which is the computationally efficient version of the DFT, are covered.

8.1 FOURIER TRANSFORMS

The Fourier transform pair for discrete aperiodic signals is given by:

Fourier transform pairs for discrete signals

$$\begin{cases} X\left(e^{j\theta}\right) = \sum_{n=-\infty}^{\infty} x[n]e^{-jn\theta}, \quad \theta = \omega T_s \\ x[n] = \frac{1}{2\pi} \int_{-\pi}^{\pi} X\left(e^{j\theta}\right) e^{jn\theta} d\theta. \end{cases} \tag{8.1}$$

These two equations allow the transformation of signals from the time to the frequency and from the frequency back to the time domain.

8.1.1 DISCRETE FOURIER TRANSFORM

Fourier transform of discrete signals is continuous over the frequency range $0–f_s/2$. Thus, from the viewpoint of programming, this transform is difficult to implement due to the integration involved. In practice, DFT is used in place of Fourier transform. DFT is the equivalent of Fourier series in the analog domain. However, it should be noted that DFT and Fourier series pairs are defined for periodic signals. These transform pairs are expressed as:

Fourier series for periodic analog signals

$$\begin{cases} X_k = \frac{1}{T} \int_{-T/2}^{T/2} x(t)e^{-j\omega_0 kt} dt \\ x(t) = \sum_{k=-\infty}^{\infty} X_k e^{j\omega_0 kt} \\ \text{where } T \text{ denotes period} \\ \text{and } \omega_0 \text{ fundamental frequency.} \end{cases} \tag{8.2}$$

Discrete Fourier transform (DFT) for periodic discrete signals

$$\begin{cases} X[k] = \sum_{n=0}^{N-1} x[n]e^{-j\frac{2\pi}{N}nk}, \; k = 0, 1, \ldots, N-1 \\ x[n] = \frac{1}{N} \sum_{k=0}^{N-1} X[k]e^{j\frac{2\pi}{N}nk}, \; n = 0, 1, \ldots, N-1. \end{cases} \tag{8.3}$$

The equation describing the DFT transformation can be written as:

$$X[k] = \sum_{n=0}^{N-1} x[n] * W_N^{nk}, \quad k = 0, 1, \ldots, N-1, \tag{8.4}$$

where $W_N = e^{-j2\pi/N}$. To compute each term, N complex multiplications and $N-1$ complex additions are required. For a frame consisting of N input samples, N^2 complex multiplications and $N^2 - N$ complex additions are thus required. It is easy to see that this method is computationally inefficient, in particular when N increases. Here is a typical DFT code in C that appears in [1]:

```c
void DFT(Complex* data){
    int i, j;
    int N = data->N;
    float arg, wI, wR;

    float sumXr[N];
    float sumXi[N];

    for (i=0; i<N; i++) {
        sumXr[i] = 0.0f;
        sumXi[i] = 0.0f;

        for(j=0; j<N; j++) {
            arg = 2*PI*i*j/N;
            wI = cos(arg);
            wR = sin(arg);
            sumXr[i] += data->real[j] * wR +
                    data->imaginary[j] * wI;
            sumXi[i] += data->imaginary[j] * wR -
                    data->real[j] * wI;
        }

    }

    for(i=0; i<N; i++) {
        out->real[i] = sumXr[i];
        out->imaginary[i] = sumXi[i];
    }
}
```

This code takes an input structure containing arrays for the real and imaginary components of a signal segment along with the number of samples contained in the signal segment. The DFT is then computed and the input signal is overwritten by the DFT. Notice that this code is computationally inefficient, as it calculates each twiddle factor (wR and wI) using a math library at every iteration. It is also important to note that the frequency-domain resolution of the DFT may be increased by increasing the size of the transform and zero-padding the input signal. Zero-padding allows representing the signal spectrum with a greater number of frequency bins. In the above code, the size of the transform matches the size of the array provided by the input structure. For example, if the input array is allowed to store four times the length of the original signal with the remaining three-quarters being zero-padded, the resulting transform will contain four times the frequency resolution of the original transform.

8.1.2   FAST FOURIER TRANSFORM

The computational complexity (number of additions and multiplications) of DFT is reduced to $(N/2) \log_2 N$ complex multiplications and $N \log_2 N$ complex additions by using FFT algorithms. In these algorithms, N is normally considered to be a power of two. To improve the computational efficiency, the FFT computation utilizes the symmetry properties in the DFT transformation. A typical FFT C code appears below (note that there are many FFT versions, here the FFT implementation in [1] is stated):

```
void FFT(Complex* data){
    int i, j, k, L, m, n, o, p, q, r;
    float tempReal, tempImaginary, cos, sin, xt, yt, arg;
    k = data->N;
    j = 0;
    m = k/2;
    float cosine[k];
    float sine[k];

    for (i=0; i<k/2; i++) {
        arg = -2*M_PI*i/k;
        cosine[i] = cos(arg);
        sine[i] = sin(arg);
    {

    //bit reversal
    for(i=1;i<(k-1);i++) {
        L=m;
```

```
        while(j>=L) {
                j = j-L;
                L = L/2;
        }


        j = j+L;

        if(i<j) {
                tempReal = data->real[i];
                tempImaginary = data->imaginary[i];
                data->real[i] = data->real[j];
                data->imaginary[i] = data->imaginary[j];
                data->real[j] = tempReal;
                data->imaginary[j] = tempImaginary;
        }
}

L = 0;
m = 1;
n = k/2;

//computation
for(i=k; i>1; i=(i>>1)) {
    L = m;
    m = 2*m;
    o = 0;

    for(j=0; j<L; j++) {
        cos = cosine[o];
        sin = sine[o];
        o = o+n;

        for(p=j; p<k; p=p+m) {
            q = p+L;

            xt = cos*data->real[q] - sin*data->imaginary[q];
            yt = sin*data->real[q] + cos*data->imaginary[q];
            data->real[q] = data->real[p] - xt;
```

```
                    data->real[p] = data->real[p] + xt;
                    data->imaginary[q] = data->imaginary[p] - yt;
                    data->imaginary[p] = data->imaginary[p] + yt;
                }
            }
            n = n>>1;
        }
    }
```

Note that, unlike in the DFT code, the twiddle factors (cos and sin) are pre-computed and stored in the cosine and sine lookup tables, respectively. This function takes the same input structure as the DFT function and performs a transformation which overwrites the input signal with the transformed signal.

8.2 LEAKAGE

When computing DFT, it is required to assume periodicity with a period of N_s samples. Figure 8.1 illustrates a sampled sinusoid which is no longer periodic. In order to make sure that the sampled version remains periodic, the analog frequency should satisfy this condition [1]:

$$f_o = \frac{m}{N_s} f_s, \tag{8.5}$$

where m denotes number of cycles over which DFT is computed.

When the periodicity constraint is not met, a phenomenon known as leakage occurs. Figure 8.2 shows the effect of leakage on the FFT computation. In this figure, the FFTs of two sinusoids with frequencies of 250 Hz and 251 Hz are shown. The amplitudes of the sinusoids are unity. Although there is only a 1 Hz difference between the sinusoids, the FFT outcomes are significantly different due to improper sampling. In Figure 8.2a, it can be seen that the signal energy resides primarily in the 250 Hz band. Leakage causes the signal energy to be spread out to the other bands of the transform. This is evident in Figure 8.2b by the diminished peak at 250 Hz and increased amplitude of the bands to either side of the peak.

8.3 WINDOWING

Leakage can be reduced by applying a windowing function to the incoming signal. In the time domain, a windowing function is shaped such that when it is applied to a signal, the beginning and end taper toward zero. One such window is the Hanning window, shown in Figure 8.3.

An example C code which generates the Hanning window appears below:

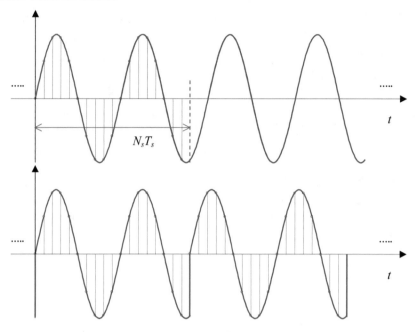

Figure 8.1: Periodicity condition of sampling.

```
float* Hanning(int N) {

    float* window = (float*)malloc(N*sizeof(float));
    for(i=0; i<N; i++) {
        window[i] = (float)((1.0-cos(2.0*M_PI*(i+1)/(N+1)))*0.5);
    }
    return window;

}
```

8.4   OVERLAP PROCESSING

It should be noted that windowing introduces distortion of the signal information. Overlap processing is often used to reduce this distortion. Overlap processing is a method by which the time-domain resolution of the Fourier transform is increased. In overlap processing, instead of processing signal samples in discrete chunks, samples are buffered and shifted through a time-domain window. Each shift through the buffer retains some of the previous signal information, on which the windowing function is applied, as illustrated in Figure 8.4. In this figure, the input signal is $x(n) = u(n) - u(n - 221)$ and a Hanning window is generated by calling

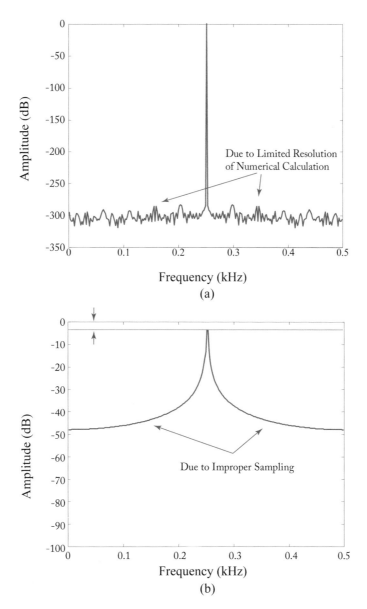

Figure 8.2: FFTs of (a) 250 Hz and (b) 251 Hz sinusoids.

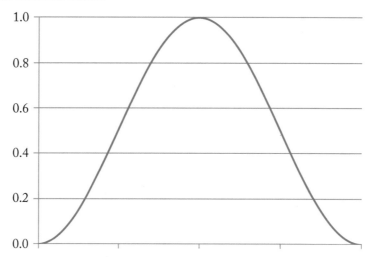

Figure 8.3: Hanning window.

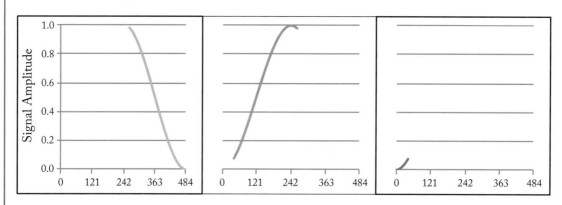

Figure 8.4: Fourier transform windowing (from left to right: iteration 1, 2, and 3).

`Hanning(485)` using the C function provided earlier. The frame size, or shift, is considered to be 221 samples which corresponds to the length of a rectangular pulse. This leads to gaining greater resolution in the time-domain.

8.5   RECONSTRUCTION

Reconstruction or synthesis is the process by which a time-domain signal is recovered from a frequency domain signal. It involves performing inverse Fourier transform and overlap-add reconstruction when overlap processing is performed.

8.5.1   INVERSE FOURIER TRANSFORM

The inverse Fourier transform is operationally very similar to the forward Fourier transform. From Eq. (8.3), one can see that to recover the time domain signal, the complex conjugate of the twiddle factor W_N can be used while scaling the resulting value by the inverse of the transform size. The code is easily implemented by modifying the code stated earlier for DFT/FFT. An inverse transform C code is shown below:

```
void iDFT(Complex* data){
      int i, j;
      int N = data->N;
      float arg, wI, wR;

      float sumXr[N];
      float sumXi[N];

      for (i=0; i<N; i++) {
            sumXr[i] = 0.0f;
            sumXi[i] = 0.0f;

            for(j=0; j<N; j++) {
                  arg = 2*PI*i*j/N;
                  wI = cos(arg);
                  wR = -sin(arg);
                  sumXr[i] += data->real[j] * wR +
                              data->imaginary[j] * wI;
                  sumXi[i] += data->imaginary[j] * wR -
                              data->real[j] * wI;
            }
      }

      for(i=0; i<N; i++) {
            out->real[i] = sumXr[i]/N;
            out->imaginary[i] = sumXi[i]/N;
      }
}
```

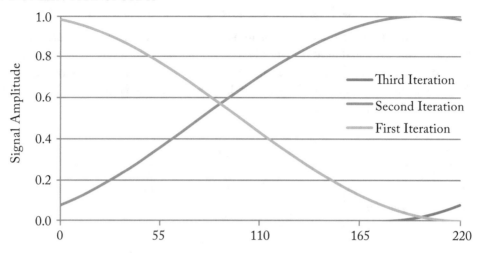

Figure 8.5: Overlap-add of analysis window.

8.5.2 OVERLAP-ADD RECONSTRUCTION

An overlap-add reconstruction is necessary when overlap processing is performed. Once the inverse Fourier transform is computed, this reconstruction involves adding together appropriate sections or segments of the signal. Continuing with the example stated earlier, i.e., $x(n) = u(n) - u(n - 221)$, and the analysis window `Hanning(485)`, the segments of the windowing function are overlaid according to the overlap-add frame size of 221 samples, shown in Figure 8.5, with the summation result shown in Figure 8.6. It can easily be seen that the amplitude of the result is not unity but has a periodic modulation with the period being equal to the frame size of 221 samples. Thus, if uncorrected, the output signal would be amplitude modulated with the fundamental frequency of $f_m = f_s/m$, where m is the frame size. This modulation can be removed by modulating the output signal of the overlap-add reconstruction with the inverse signal, shown in Figure 8.7. The final result will be an output signal with the correct amplitude.

L8 LAB 8: FREQUENCY DOMAIN TRANSFORMS – DFT AND FFT

In this lab, the C implementations of DFT and FFT are considered.

In the previous filtering labs, although audio data samples were passed in frames (in order to accommodate the requirements of the audio APIs), the actual filtering operation was done on a sample by sample basis using linear convolution. However, in performing DFT (or FFT), the transform requires access to a window of audio data samples which may or may not contain more than one frame of data. This is referred to as frame processing. In frame processing, N samples

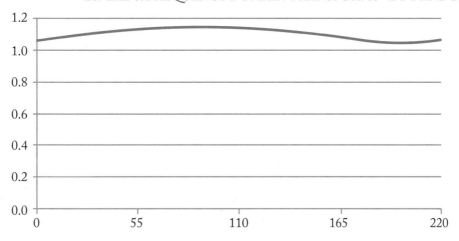

Figure 8.6: **Analysis window summation.**

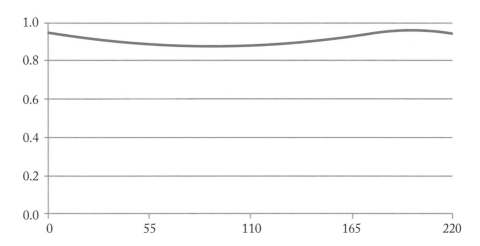

Figure 8.7: **Synthesis window.**

need to be captured first and then operations are applied to all N samples with the computation time measured in terms of the duration of a frame.

The application shell for this lab resembles that of the previous labs. The code basically follows the same initialization, computation, and finalization methods covered in the previous labs. The DFT and FFT implementations, given in Chapter 8, appear in the file Transforms.c. In addition, an audio spectrogram application is provided for the Android and iOS targets by taking into consideration the concepts discussed in the chapter. This app features graphical display of

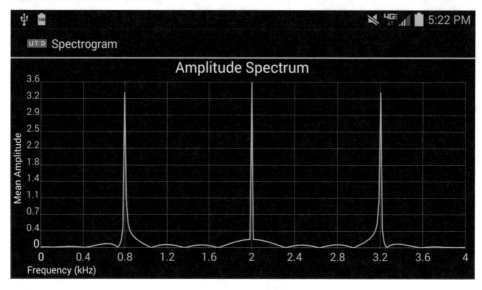

Figure 8.8: Android spectrogram app graphical display.

the frequency spectrum and allows the adjustment of the transform parameters (see Figures 8.8–8.10).

Instructions for running the Spectrogram lab based on XCode as follows.

- In the project, open the Podfile and change iOS to a new version if needed.

- Steps for the installation of Ruby and CocoaPods versions:

  Get Ruby 2.6.3 and use Ruby Version Manager to install a new version of Ruby by taking the following steps:

Open *Terminal* and *Run*:

```
curl -L https://get.rvm.io| bash -s stable
```

Then, reopen *Terminal* and *Run*:

```
rvm install ruby-2.6

rvm use ruby-2.6.3

rvm---default use 2.6.3
```

Install an older version of CocoaPods using the following code on the *Terminal* :

Figure 8.9: Android spectrogram app main screen and settings menu.

```
sudo gem install cocoapods -v 1.8.4
```

Next, open *Terminal* and change the directory to the project folder and *Run*:

```
pod install
```

- In the project, open Spectrogram.xcworkspace (not Spectrogram.xcodeproj) to incorporate the plotting library.

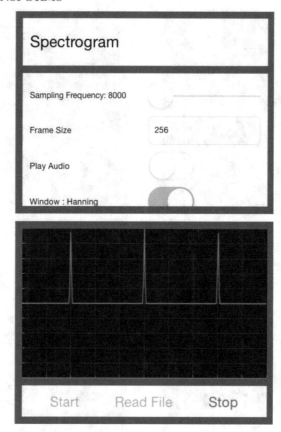

Figure 8.10: iOS spectrogram app graphical display.

L8.1   LAB EXERCISES

1. **Computational Complexity:** Compare the processing time of the DFT vs. FFT implementation of the Fourier transform. Does the DFT meet real-time performance constraints? How does the processing time of the two implementations change as the transform size is increased?

2. **Linear Convolution:** Consider a linear time invariant system which comprises a bandpass filter with the following specifications:

   - 8000 Hz sampling rate
   - −90 dB stopband attenuation
   - 800–1800 Hz passband
   - 100 Hz wide transition bands

Find the output of the system $y(n)$ to an input audio signal $x(n)$ via the overlap and add convolution method. Record the processing time for the case with 256-sample frames as input. It helps to use the MATLAB *fdesign* tool to generate the filter coefficients for this system.

3. **Frequency Domain Filtering:** Solve the previous bandpass filter system in the frequency domain by using two forward and one inverse FFT by using $Y(k) = H(k)X(k)$ (convolution property). For the frequency domain case, consider the results when using 512-point FFTs and 256 sample frames for the following two cases:

   - overlap processing with 50% overlap and Hanning windowing; and
   - no overlap and rectangular windowing.

Use the test signals provided to examine both of the filter implementations. Once the frequency domain filtering is completed, copy back the debugged outputs and compare all three filtering outputs using MATLAB. Note that the frequency domain representation may be stored between calls and does not need to be re-computed after the initial computation.

4. **Reconstruction:** Using the test signals provided, how do parameter adjustments such as the transform size, windowing function, window size, and overlap affect the spectrum? Examine how these parameters affect the audio signal when it is synthesized from its frequency spectrum.

8.7   REFERENCES

[1] *TI Application Report SPRA291*. http://www.ti.com/lit/an/spra291/spra291.pdf   122, 123, 125, 142

CHAPTER 9

Code Optimization

In this chapter, code optimization techniques which often have a major impact on the computational efficiency of C codes are covered. These techniques include compiler optimizations, efficient C code writing, and architecture-specific instructions of the ARM processor. For a better understanding of these techniques, they are illustrated through the signal processing example app of linear convolution. In general, to write an efficient C code, it helps to know how the processor implements it the way it is written.

9.1 CODE TIMING

When attempting to increase code efficiency, it is of great help to find out which segment or segments are running inefficiently. This can be achieved by manually timing code segment or segments via a high-resolution system clock. An imprecise way of timing is a time display of the computer being used. A more precise way is to use C clock functions in the header file *time.h* (Android) or *mach_time.h* (iOS). On Android targets, the function *clock_gettime* with the time sources CLOCK_REALTIME and CLOCK_MONOTONIC (utilizing Java clock functions of *System.currentTimeMillis()* and *System.nanoTime()*) provide timing functionality. The example code segments below indicate how *time.h* is used for this purpose. For Android:

```
struct timespec startTime, stopTime;
clock_gettime(CLOCK_MONOTONIC, &startTime);
/*
      ---Code to be analyzed---
*/
clock_gettime(CLOCK_MONOTONIC, &stopTime);
unsigned long long totalTime = (stopTime.tv_sec - startTime.tv_sec)
      * 1000000000LL + stopTime.tv_nsec - startTime.tv_nsec;
//totalTime contains time elapsed in nanoseconds
```

and for iOS:

```
clock_t start, finish;
start = clock();
```

```
/*
        ---Code to be analyzed---
*/
finish = clock();
double elapsedTime = (double)(finish---start)/CLOCKS_PER_SEC;
//elapsedTime contains time elapsed in seconds
```

The variables will show the total execution time of any code placed where the comment section is indicated. More information on timing functionality may be found in the documentation for the relevant C headers.

9.2   LINEAR CONVOLUTION

The linear convolution (FIR filtering) operation provides a good example of how a signal processing C code should be written to be computationally efficient. For portability purposes, the FIR filtering C code, which appears below, is implemented by setting up variable storage during initialization and by generating output when the compute method is called.

```
typedef struct FIRFilter {  // Data storage structure
        int numCoefficients;
        int frameSize;
        float* coefficients;
        float* window;
        float* result;
} FIRFilter;

FIRFilter* //Data structure initialization
newFIR(int frameSize, int numCoefficients, float* coefficients) {
    FIRFilter* newFIR = (FIRFilter*)malloc(sizeof(FIRFilter));

    newFIR->numCoefficients = numCoefficients;
    newFIR->frameSize = frameSize;
    newFIR->coefficients = (float*)malloc(numCoefficients*
                                            sizeof(float));
    newFIR->window = (float*)calloc(numCoefficients + frameSize,
                                            sizeof(float));
    newFIR->result = (float*)malloc(frameSize*sizeof(float));
    int i;
```

```
        for(i=0;i < numCoefficients;i++) {
            newFIR->coefficients[(numCoefficients - (1) - i] =
                                (float)coefficients[i];
        }
        return newFIR;
    }

void computeFIR(FIRFilter* fir, float* input) {
        int i, j;
        float temp;

        for(i=0; i<fir->numCoefficients; i++) {
            fir->window[i] = fir->window[fir->frameSize + i];
        }

        for(i=0; i<fir->frameSize; i++) {
            temp = 0;
            fir->window[fir->numCoefficients + i] = input[i];
            for(j=0; j<fir->numCoefficients; j++) {
                temp += fir->window[i + j + 1] * fir->coefficients[j];
            }
            fir->result[i] = temp;
        }
    }
```

Note that during the initialization, the array describing the coefficients which will be applied during the linear convolution are being flipped into reverse order so that access to the window and coefficients arrays are made using increasing address indices. The data type conversion from short to floating-point is done elsewhere.

9.3   COMPILER OPTIONS

The simplest but an effective optimization step involves specifying the optimization options provided by the compiler. The Android NDK toolchain and Xcode IDE are based on the GCC compiler. Thus, the optimization options provided by GCC can be used when generating native libraries for Android and iOS apps. A complete list of the optimization options can be found on the GNU GCC documentation website:

http://gcc.gnu.org/onlinedocs/gcc-4.6.0/gcc/Optimize-Options.html

The compiler allows applying different optimization levels. These levels refer to different optimization flags that the builder enables, and are explained in detail at the above link. The optimization levels considered here are −O{0-3}, −Os, and −Ofast. The builder asserts many compilation options by default; among these is the −Os flag which denotes optimization for code size. The −O0 option denotes no optimization flags; −O1 enables a subset of options; −O2 enables more options adding to the ones enabled by −O1; and −O3 includes all the optimizations added by −O1 and −O2. The option −Ofast enables optimizations that may result in variables getting truncated or rounded incorrectly for floating-point math operations. For most cases, the −O3 option produces the best computational efficiency outcome.

When using Android Studio, the options for C code libraries need to be set using the *build.gradle* file of the app. The optimization flags can be set within the *ndk* block using the *cFlags* directive. An example using the −O3 optimization follows:

```
ndk {
        moduleName ''yourLibrary''
        abiFilter ''armeabi''
        ldLibs ''log''
        cFlags ''-O3''
}
```

When using Xcode, all options for C code libraries can be set within the Build Settings of the app by changing the Optimization Level under the Apple LLVM 6.1—Code Generation section.

9.4   EFFICIENT C CODE WRITING

The compiler automatically performs common code optimization changes, such as loop reversal or changing division by a constant to multiplication by the reciprocal of the constant. Thus, it may only be necessary to further improve code efficiency by refactoring or manually implementing architecture specific features such as SIMD instructions. Let us examine the changes that can be made to the above linear convolution code to improve its computational efficiency or performance.

For the FIR filter to work properly, it is required to store a sufficient number of previous input samples in memory. Because the generic ARM processor does not support circular buffering, this can be accomplished by using two loops to shift previous samples through an array structure in memory as follows:

```
for(i=0; i<fir->numCoefficients; i++) {
        fir->window[i] = fir->window[fir->frameSize + i];
```

```
}

for(i=0; i<fir->frameSize; i++) {
      fir->window[fir->numCoefficients + i] = input[i];
}
```

The array `window` is stored in heap memory using the previously defined FIRFilter structure as these values need to be retained between calls to the compute method. Memory allocation is time consuming and multiple repeated allocations should be avoided if possible.

Another way to improve code performance is to reduce the logic necessary for the loop to operate. Although the above two loops may appear fine, it still takes extra operations to compute the array index and thus the memory address of the desired value. A method involving pointer manipulation can be used as shown in the following code block:

```
void computeFIR(FIRFilter* fir, float* input) {
      int i, j;
      float temp;
      float* windowPtr = fir->window;

      for(i=0; i<fir->numCoefficients; i++) {
            *windowPtr = windowPtr[fir->frameSize];
            windowPtr++;
      }

      for(i=0; i<fir->frameSize; i++) {
            temp = 0;
            *windowPtr = input[i];
            for(j=0; j<fir->numCoefficients; j++) {
                  temp += windowPtr[-j] * fir->coefficients[j];
            }
            windowPtr++;
            fir->result[i] = temp;
      }
}
```

Using this technique, the memory address of the array is loaded one time before variable overwrites or computations take place. Coming out of the shifting loop, the pointer `windowPtr` refers to the memory location of the first array index that receives a sample from the new frame of audio data due to the post-update incrementing. Using the pointer also removes the need for some logic to accomplish array indexing. In terms of actual instructions generated by the compiler, this version of the code has six operations in the second loop as opposed to the original version of the code having ten operations. Also note, unlike the previous case where the window array was accessed from low index values to high index values, the window array is now being accessed in reverse order.

The instructions to compute the result can be generalized into core instructions, e.g., the multiply-accumulate instruction in linear convolution. Supporting instructions, which add computational overhead, are responsible for memory address generation and transformation, data loading and storage, and branch or loop logic. In comparison to the original code version, the pointer manipulation approach reduces the number of supporting instructions by lowering the overhead for memory address generation but the core computation remains the same.

Many other efficient C code writing techniques are covered in [1] and the reader is referred to this reference.

9.5   ARCHITECTURE-SPECIFIC OPTIMIZATIONS

If the performance of a code is not adequately improved by compiler optimization options or by refactoring, one may resort to using architecture-specific optimization. On ARM processors, this consists of using specialized hardware capabilities built into the processor.

9.5.1   TARGET ARCHITECTURE

It is important to compile codes for the proper hardware capabilities of the smartphone target. This subsection is focused on Android targets due to the wide range of target architectures used, whereas iOS only uses ARMv7 and ARMv8 architectures.

A target architecture can be specified by listing the desired ABIs (Application Binary Interfaces) in the file *build.gradle* as indicated below:

```
ndk {
      moduleName ''yourLibrary''
      abiFilter ''armeabi armeabi-v7a mips x86''
      ldLibs ''log''
      cFlags ''-O3''
}
```

This generates native libraries compiled specifically for the targets listed in the abiFilter directive. For each targeted ABI, the compiler generates a native library which gets included with the application. By default, all available ABIs will be built by the build system. When the application is installed, the corresponding library also gets installed.

Architecture-specific optimizations may be included by setting flags and enabling code sections at compile time. Flags can be set in the *build.gradle* file by checking the target using the productFlavors directive as follows:

```
productFlavors {
    armv7 {
        ndk {
            abiFilter ''armeabi-v7a''
            cFlags ''-mfloat-abi=softfp -mfpu=neon -march=armv7-a
                                            -DMETHOD=1''
        }
    }
}
```

Code sections can then be enabled or disabled depending on the flags set when the library is getting compiled. This allows architecture-specific optimizations to be included in one main set of source files. As noted below, the METHOD flag is defined to enable NEON code blocks:

```
#if METHOD == 1
      /* Normal code */
#elif METHOD = 2
      /* NEON code */
#endif
```

Note that this is only one case and the Gradle build system allows compilation of completely separate source sets for different architectures. This eliminates the need for using compiler flags for source code selection when building for different architectures. In addition, separate compilation flags may be set for each product flavor, allowing one to fine-tune to a specific architecture. It is to be emphasized that this discussion of the Gradle build system may change due to the relatively recent release of Android Studio as well as the continued development effort by Google on the Android Studio IDE.

9.5.2   ARM HARDWARE CAPABILITIES

Often, significant gains in performance can be acquired by enabling compilation for the hardware architecture version that is being used. For example, the compilation setting *armeabi* refers to processors up to ARMv6. When using ARMv7, the compilation setting *armeabi-v7a* provides additional instruction sets such as Thumb-2 and VFPv3 (vector floating-point). One major disadvantage of ARMv6 is the absence of a hardware floating-point unit. This results in floating-point operations to be performed via software routines instead of a dedicated hardware. ARMv7 allows hardware floating-point operations with the addition of the VFPv3 instruction set. Another feature introduced with ARMv7 is the Advanced SIMD instruction set provided by the NEON Media Processing Engine (NEON MPE) or coprocessor. These instructions are similar to the MMX and SSE SIMD (single instruction, multiple-data) instruction sets on Intel processors.

The Advanced SIMD instruction set includes many functions specifically targeted for signal processing applications. For example, in the linear convolution code, the core multiply-accumulate statement consists of a multiply operation followed by an addition operation. This causes a value getting rounded after both of the multiplication and the addition operations are completed. By using VFPv4, this can be accomplished in one fused multiply-accumulate instruction resulting in only one rounding.

9.5.3   NEON INTRINSICS

Although SIMD instructions can be used when writing hand-optimized assembly code, this approach would be cumbersome and time consuming. Fortunately, a mechanism to come close to this approach is the use of NEON intrinsics within a C code by using the functions provided in the header *arm_neon.h*. This was encountered during a previous lab involving the Newton–Raphson iteration for finding inverse and square root of a number.

However, it is important to note that the use of intrinsics has drawbacks in terms of aligned data. Aligned data refers to data stored in memory such that the base address is a multiple of powers of two, and data accesses are performed on the same data stride length. For ARM, an effective data alignment would be the one that matches the size of a cache line in level 1 cache. For example, on the Cortex-A15 ARM processor, the cache line size is 64 bytes. Aligned data loads allow the processor to read ahead in memory and load data into the level 1 cache before it is read into the register file, resulting in decreased loading times. Aligned data loads cannot be performed when using intrinsics, and memory pointer increments cannot be done as part of load or store operations. The only way to incorporate these features is via assembly code.

Now, let us consider the situation involving floating-point arithmetic. The filtering code version when using NEON intrinsics is stated below:

```
void computeFIR(FIRFilter* fir, float* input) {
    int i, j;
    float32x4_t freg1, freg2, freg3;  //temporary registers
    float32_t* coeffsPnt;  //temporary pointer to coefficients array

    //temporary pointers to window buffer
    float32_t* windowPnt1 = fir->window;
    float32_t* windowPnt2 = &(fir->window[fir->frameSize]);
```

```
//Assuming the number of coefficients is a multiple of 4
for(i=0; i<fir->numCoefficients; i+=4) {
      //load elements starting at window[numCoefficients + i]
      //and shift to window[i]
      freg1 = vld1q_f32(windowPnt2);
      windowPnt2 += 4;
      vst1q_f32(windowPnt1, freg1);
      windowPnt1 += 4;
}

for(i=0; i<fir->frameSize; i+=4) {
      freg1 = vld1q_f32(input); //load first four elements
                                                    of input
      input += 4;

      vst1q_f32(windowPnt1, freg1); //store in window buffer
      windowPnt1 += 4;
}

for(i=0; i<fir->frameSize; i++) {
      windowPnt2 = fir->window + 1 + i; //copy pointers
      coeffsPnt = fir->coefficients;

      freg3 = vdupq_n_f32(0.0);  // initialize accumulator to zero

      for(j=0; j<fir->numCoefficients; j+=4) {
            //load four elements of input
            freg1 = vld1q_f32(windowPnt2);
            windowPnt2 += 4;

            //load four filter coefficients
            freg2 = vld1q_f32(coeffsPnt);
            coeffsPnt += 4;

            //multiply-accumulate - freg3 = freg1*freg2+freg3
            freg3 = vmlaq_f32(freg3, freg1, freg2);
      }
```

```
            //save output
            fir->result[i] = (freg3[0] + freg3[1] + freg3[2] + freg3[3]);
        }
    }
}
```

The overall result is the same as the previous code versions, but now the linear convolution result is computed with vectors containing four elements each.

L9   LAB 9:
CODE OPTIMIZATION

The purpose of this lab is to experiment with the optimization steps discussed above. These steps include changing compiler settings, writing efficient code constructs, and using architecture-specific instructions for the ARM processor. The FIR filtering (linear convolution) example is considered as a model case to show the effects of these steps on the real-time throughput.

Consider a lowpass filter whose passband covers the human vocal frequency range. The specification used to generate the filter in MATLAB is as follows:

```
rpass = 0.1;                                %passband ripple
rstop = 20;                                 %stopbad ripple
fs = 48000;                                 %sampling frequency
f = [3000 3570];                            %frequency bands
a = [1 0];                                  %desired amplitudes
dev = [(10^(rpass/20)-1)/(10^(rpass/20)+1) 10^(-rstop/20)];
                                            %deviations
[n,fo,ao,w] = firpmord(f,a,dev,fs);         %estimate
B = firpm(n,fo,ao,w);                       %compute coefficients
```

The above MATLAB code produces the coefficient array *B* containing 128 coefficients. The shell for this lab provides the timing and linear convolution functions.

L9.1   COMPILER OPTIONS

Using the filter specified above and a sampling rate of 48 kHz, run the filter by enabling different optimization levels and report the processing times achieved. Use the recording function and wait until the reported frame time stabilizes, or use a sufficiently long test signal (~ 20 s) and record the processing time for the signal.

L9.2 TARGET ARCHITECTURE (ANDROID ONLY)

Using a target smartphone which supports armeabi-v7a, enable the abiFilter for armeabi-v7a in the ndk section of the build.gradle file and enable the hardware floating-point by setting the cflags to `-mfloat-abi=softfp -mfpu=neon-vfpv4 -O3` and re-run the experiment. Compare the processing time obtained with armeabi vs. armeabi-v7a.

L9.3 CODE MODIFICATION

Implement the linear convolution algorithm using the discussed pointer manipulation technique and report the processing time. Compare the processing time when using floating-point values for the filter coefficients with the processing time when using double precision format for the filter coefficients. This can be done by changing the data type of the coefficient storage array. For these experiments, use the $-O3$ compiler optimization setting.

Advanced SIMD (Floating-Point)

Implement the linear convolution algorithm by using NEON intrinsics alongside the pointer manipulation technique by using the flag METHOD described earlier. Run the code and report the processing time. Indicate whether the processing time using NEON matches with that of the other versions.

Advanced SIMD (Fixed-Point)

Use NEON to implement the filter using fixed-point arithmetic. Use Q15 format to quantize and export the filter coefficients via MATLAB. Report the processing time of this fixed-point implementation version and compare the output signal with that of the floating-point implementation version.

The following links provide information on the NEON saturating instruction type, including the instruction VQDMLAL:

http://infocenter.arm.com/help/index.jsp?topic=/com.arm.doc.dui0473c/CJAEFAIC.html
http://infocenter.arm.com/help/index.jsp?topic=/com.arm.doc.dui0489i/CIHGJIHD.html

The instruction VQDMLAL performs the operation $z = z + ((x * y) \ll 1)$, where x and y are in Q15, and z is in Q31 format. If the instruction causes overflow, the result is saturated as described in the links.

9.7 REFERENCES

[1] http://tools.android.com/tech-docs/new-build-system

[2] A. Sloss, D. Symes, and C. Wright, *ARM System Developer's Guide*, Morgan Kaufmann Publishers, 2004. DOI: 10.1016/b978-1-55860-874-0.x5000-x.

CHAPTER 10

Implementation via MATLAB Coder

This chapter presents the steps one needs to take in order to run a signal processing algorithm written in MATLAB on the ARM processor of smartphones which were initially reported in [1] by using Simulink. The steps needed are best conveyed by going through an example. This example involves the linear convolution filtering algorithm. Considering that MATLAB programming is widely used in signal processing, the approach presented in this chapter allows running on smartphones many signal processing algorithms which are already written in MATLAB and publicly available.

10.1 MATLAB FUNCTION DESIGN

This section provides the guidelines for implementing the linear convolution filtering algorithm via a MATLAB script. The first step is to open MATLAB and create a new function file in which to implement the algorithm. The following example code provides an implementation of a frame-based finite impulse response filter (coefficients omitted). Of particular importance in this function is the usage of the persistent variable `buffer`. This variable stores previous samples of the input signal between calls to the `FIR` function so that the proper filter output is produced. The actual filtering result is computed using the built-in MATLAB `filter` function.

```
function output = FIR(input)
coefficients = []; %FIR filter coefficients

persistent buffer;
if isempty(buffer)
    buffer = zeros(1, size(input, (2)~+ size(coefficients, 2));
end
buffer = [buffer(:, end-size(coefficients,2)+1 : end) input];

filtered = filter(coefficients, 1, buffer);
output = filtered( : , size(coefficients, (2)~+ 1 : end);]
```

10.2   TEST BENCH

An important step in implementing any signal processing algorithm is functional verification. This has benefits both for debugging purposes as well as simulating the response on a target platform on which the algorithm is to run. For audio signal processing, MATLAB assumes samples of an entire audio signal are available but on an actual target, audio signal processing is done in a frame-based manner. Thus, one needs to implement frame-based processing. The following script shows how such an implementation is achieved.

```
clear;
sampleRate=8000;
sampleTime=1/sampleRate;
frameSize = 256;
time=[0:sampleTime:10*sampleRate*sampleTime];

%generate signal
length = floor(size(time,2)/frameSize)*frameSize;
time = time(1:length);
signal = 1/2*square(2*pi*600*time,50);
audiowrite('Square_duty-50_600Hz_10s.wav',signal,sampleRate);

%simulate frame-based processing
signal = reshape(signal,frameSize,[]);
signal = signal';

result = zeros(size(signal));
for i = 1:size(signal,1)
    result(i,:) = FIR(signal(i,:));
end
```

This script generates a test signal and writes it to a file for use on a target platform. The signal is reshaped into a matrix of frame-sized columns and transposed to form frame-sized rows as would be the behavior on the target. The rows are then passed to the MATLAB function for processing.

10.3   CODE GENERATION

The above MATLAB function and the test bench script steps allow generating a C code using the MATLAB Coder. The MATLAB Coder (Coder) [2] can be found in the Apps section of the toolbar. The following screenshot shows the process of using a MATLAB function to generate a C source code.

Figure 10.1: MATLAB coder function selection.

Figure 10.1 shows the initial screen for the Coder. Start by selecting the function to be converted and change the Numeric Conversion option to single precision floating-point arithmetic. In Figure 10.1, if the *Numeric Conversion* entry is not visible, this means that not all the MATLAB packages were not installed. To correct this, uninstall and reinstall MATLAB and select all the packages.

After the function is selected, the input types need to be specified (see Figure 10.2). This can be done manually or by using the above test bench script to automatically determine the data types. Select the test bench script and choose the Autodefine Input Types option to complete this step.

After the input types are set, the Coder then checks to ensure that it is able to generate a C code from the provided MATLAB script. Figure 10.3 shows the outcome with no detected errors. Although many of the built-in MATLAB functions work with the MATLAB Coder,

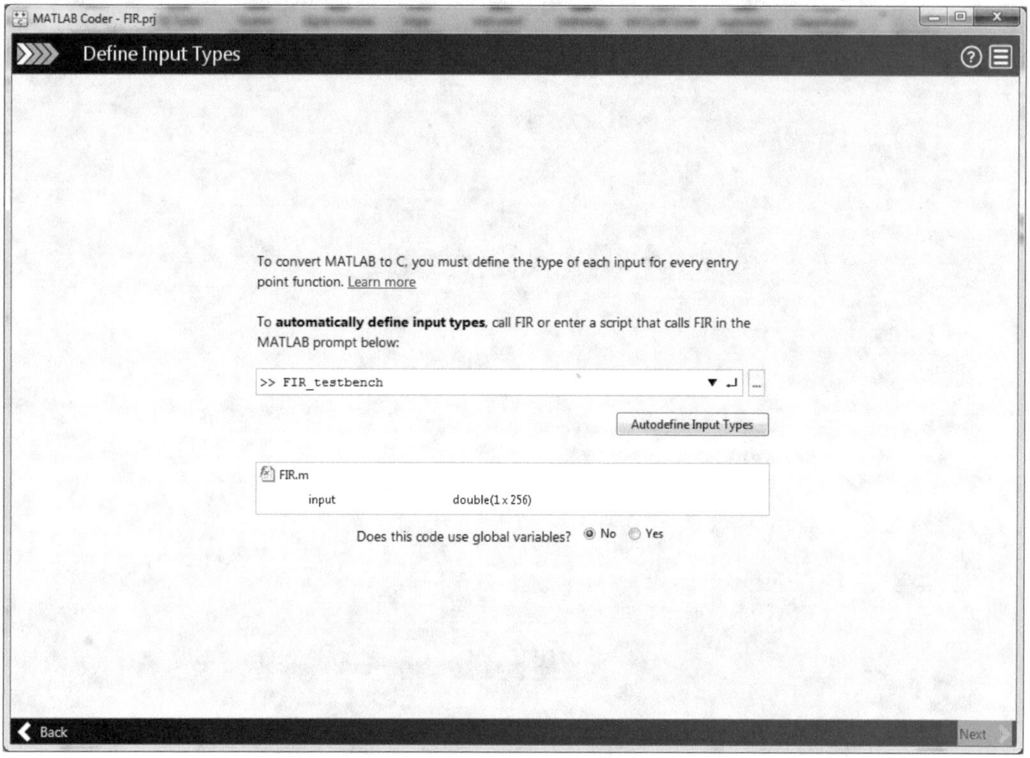

Figure 10.2: Input type specification.

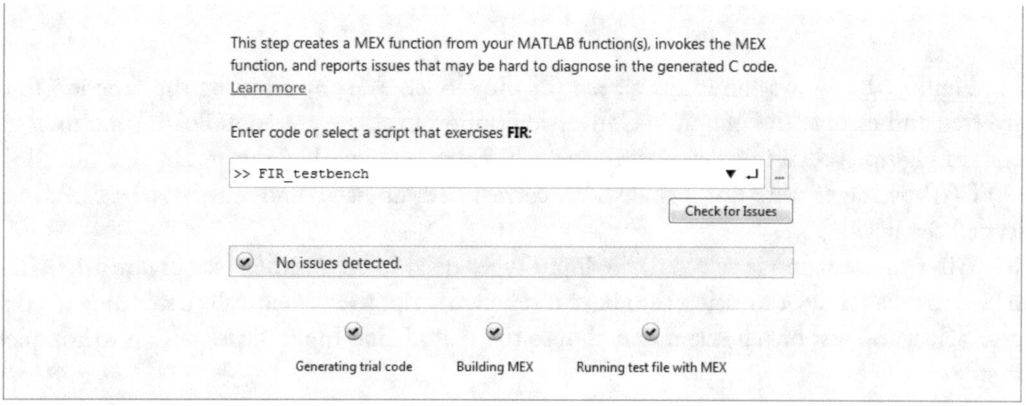

Figure 10.3: Function error check.

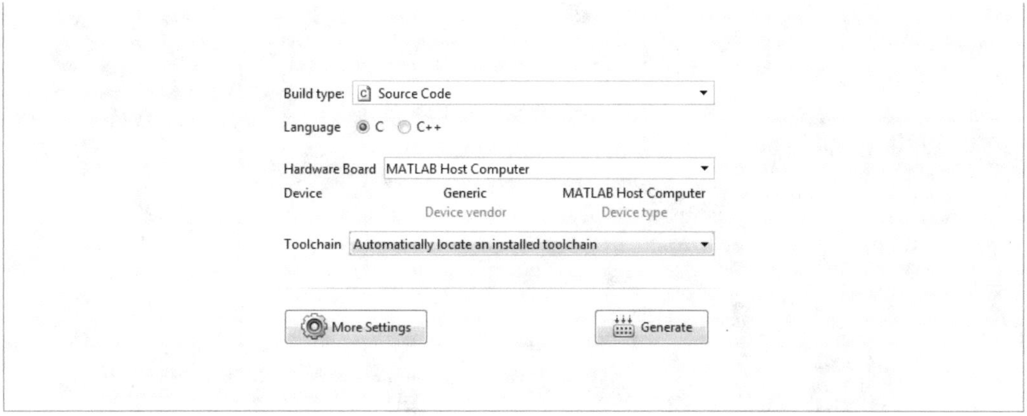

Figure 10.4: C Source Generation.

not all are supported. Unsupported functions are required to be written from scratch by the programmer.

Once the MATLAB script is checked and passed, a corresponding C source code can be generated by pressing the Generate button (see Figure 10.4). Although various configuration settings are available, the default settings are adequate for the generation of C source codes.

10.4   SOURCE CODE INTEGRATION

The final step to implement a signal processing algorithm written in MATLAB on smartphones is to deploy the generated C code on a suitable target device. For this step, an application shell is provided here in which the generated C source code needs to be placed. This shell operates in the same manner as the test bench script stated earlier; that is audio signal samples may be read from a file or recorded using the target microphone, then data samples are passed to the processing code in frames and are returned in frames. Figure 10.5 shows how the generated C code is integrated into the Android shell provided and Figure 10.6 shows how the generated C code is integrated into the iOS shell provided. If using MATLAB R2016b or later, it is required that this specific header file *tmwtypes.h* is added into the *jni* folder separately. This header file can be found in the MATLAB root with this path `MATLAB root\R2019b\extern\include`. After placing the generated C code in the *jni* folder and cleaning the project, if an error related to *omp_init_nest_lock(&emlrtNestLockGlobal)* is encountered, open the C files of *initialize* and *terminate* in the *jni* folder and comment out the following line as follows:

```
//omp_init_nest_lock(&emlrtNestLockGlobal);
```

Figure 10.5: C source integration Android.

The MATLAB Coder produces C codes with the necessary include statements and function calls for using them. One minor modification of a generated C code that may be required is to ensure having the correct input and output variable data types. In the case of array inputs, the generated code is specified using static array sizes and thus needs to be modified to access array pointers.

10.5 SUMMARY

MATLAB is extensively used to implement signal processing algorithms. The most important consideration when transiting a MATLAB function to smartphones is awareness of input and output data types. A persistent variable storage needs to be established by declaring a persistent variable and performing a one-time initialization. After this declaration and initialization, any data may be retained between calls to the function. With the approach presented, practically any signal processing algorithm written in MATLAB can be compiled to run on Android or iOS smartphones.

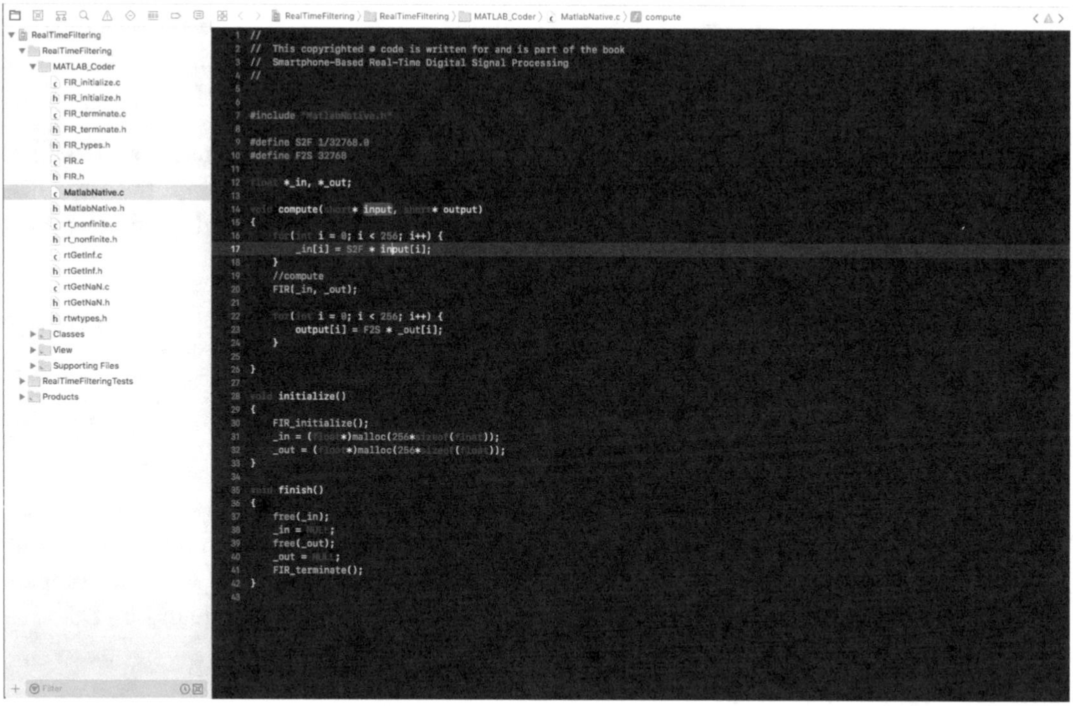

Figure 10.6: C source integration iOS.

L10   LAB 10: MATLAB CODER IMPLEMENTATION

The purpose of this lab is to transition a signal processing algorithm written in MATLAB to smartphones by generating a C source code using the MATLAB Coder and then running the C code on a smartphone ARM processor. The application shells provided in the previous labs are used here for collecting audio signal samples, passing frames of signal samples to the C code for processing, and saving the output to a file for analysis.

L10.1   LAB EXERCISES

1. Use MATLAB to design an FIR filter based on the specifications noted below using the *fir2* function.

```
f = [0 300 600 1400 2000 4000];    %frequencies (0 to nyquist)
a = [0 .33 .89 .25 .60 0];         %desired amplitude response
order = 31;                        %filter order N
```

Follow the process outlined in the chapter and integrate these filter coefficients into the MATLAB function. Use the MATLAB Coder to generate a C code, then deploy it onto a smartphone target and verify its correct functionality.

2. Write a MATLAB function to implement an adaptive FIR filter to match an IIR filter using the *yulewalk* function and the following specifications. Go through the steps discussed in the chapter to run this function on a smartphone target and verify its correct functionality.

```
Nc=8;                          %8th order IIR
f=[0 .25 .33 .66 .67 1];       %frequency band definition
m=[0 .50 .20 1 0 0];           %gain definition
```

3. Write a MATLAB function to perform frequency-domain filtering and signal reconstruction. Go through the steps discussed in the chapter to run this function on a smartphone target and verify its correct functionality. Use the same specifications as Lab 8 and compare the processing time and the result with that of the implementation done in Lab L8.

10.7   REFERENCES

[1] R. Pourreza-Shahri, S. Parris, F. Saki, I. Panahi, and N. Kehtarnavaz, From simulink to smartphone: Signal processing application examples, *Proc. of IEEE ICASSP Conference*, Australia, April 2015. 149

[2] https://www.mathworks.com/products/matlab-coder.html 150

Authors' Biographies

NASSER KEHTARNAVAZ

Nasser Kehtarnavaz is an Erik Jonsson Distinguished Professor in the Department of Electrical and Computer Engineering at the University of Texas at Dallas. His research areas include signal and image processing, real-time implementation on embedded processors, and deep learning. He has authored or co-authored more than 400 publications and 9 other books pertaining to signal and image processing, and regularly teaches an applied digital signal processing course, for which this book was developed. Dr. Kehtarnavaz is a Fellow of IEEE, a Fellow of SPIE, and a licensed Professional Engineer. www.utdallas.edu/~kehtar.

ABHISHEK SEHGAL

Abhishek Sehgal is a Senior Research Engineer at Samsung Research America. He received his B.E. degree in Instrumentation Technology from Visvesvaraya Technological University in India in 2012, and his M.S. and Ph.D. degrees in Electrical Engineering from the University of Texas at Dallas in 2015 and 2019, respectively. His research interests include signal and image processing, and real-time implementation of signal and image processing algorithms.

SHANE PARRIS

Shane Parris is a Software Engineer at Ford Motor Company. He received his B.S. degree in Electrical Engineering from the University of Texas at Dallas in 2013. His research interests include signal and image processing, and real-time implementation of signal and image processing algorithms.

ARIAN AZARANG

Arian Azarang is a Ph.D. candidate in the Department of Electrical and Computer Engineering at the University of Texas at Dallas. His research interests include signal and image processing, deep learning, remote sensing, and chaos theory. He has authored or co-authored 14 publications in these areas.

Index

Printed in the United States
by Baker & Taylor Publisher Services